CATALOGUE OF
BRITISH AND FOREIGN
War Medals
CROSSES,
BADGES, DECORATIONS,
AND MISCELLANEOUS MEDALS
IN THE COLLECTION OF
LIEUT.-COL. H. F. EATON

The Naval & Military Press Ltd

published in association with

FIREPOWER
The Royal Artillery Museum
Woolwich

Published by

The Naval & Military Press Ltd

Unit 10 Ridgewood Industrial Park,

Uckfield, East Sussex,

TN22 5QE England

Tel: +44 (0) 1825 749494

Fax: +44 (0) 1825 765701

www.naval-military-press.com

in association with

FIREPOWER

The Royal Artillery Museum, Woolwich

www.firepower.org.uk

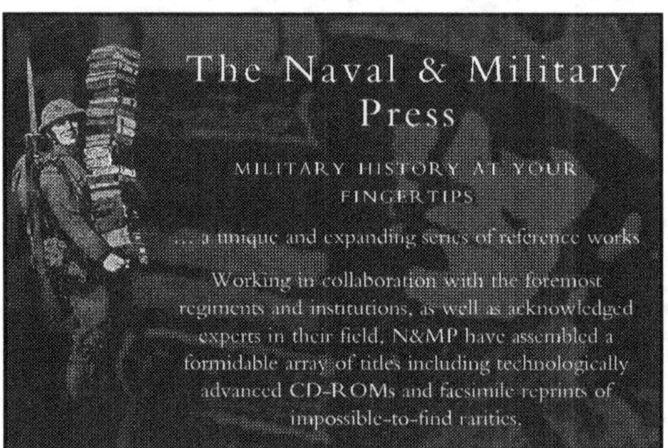

*In reprinting in facsimile from the original, any imperfections are inevitably reproduced
and the quality may fall short of modern type and cartographic standards.*

PREFACE.

THE following Catalogue has been compiled as an Index to the different Medals in this Collection, and not in any sense as a personal history of the wearers of the decorations, or of the circumstances under which they have been earned.

The original object in collecting Medals was to prevent the entire disappearance of honourable decorations granted for good service, which might occur through death of owner, accidental loss, sale, etc., etc. ; and having purchased many with this view, I conceived the design of collecting such a number of them as would illustrate a history of the War Medals of this country, from 1650 to the present time.

While engaged in this occupation, much interesting information was necessarily brought to my knowledge in respect of the individuals whose Medals I had become possessed of ; but, after much thought, I have determined to keep the Catalogue as concise as possible, and only to insert in it such matter as will serve to classify the different specimens, and so to render the Collection intelligible to such persons as may be interested in them.

It is hoped that, in addition to the above, the Catalogue may serve a yet higher purpose, in reminding the reader that brave men are found in all ranks of the Army and Navy, and this Collection is a sign that their services are recognised by their country.

H. F. EATON,
CAPT. AND LIEUT.-COL. GREN. GUARDS.

May, 1880.

CONTENTS.

CASE I.

WAR MEDALS, 1793 TO 1814.

GOLD MEDAL.

Nicolas' " Orders of Knighthood,"

Vol. IV., Page 32.

" The glorious frequency of victories in the Peninsula during the years 1808, 1809, caused two gold medals to be instituted for the reward of such superior officers as had distinguished themselves; and the same decorations were subsequently conferred for military services against the enemy in other parts of the world.

" The larger is $1\frac{1}{2}$ inch in diameter, and has on the obverse, Britannia, wearing a helmet and seated on the globe, her right hand extended, holding a wreath of laurel; in her left is a palm branch. To her right is the lion of England; and, on the left, a round shield charged with the crosses of the Union banner. The reverse bears a wreath of laurel, within which the name of the event is inscribed, and the year. On the edge, the name and rank of the officer to whom the medal was given, are engraved. The large medal was reserved for general officers, and, in full dress, is worn from the neck to a crimson riband with blue edges.

" The smaller medal is $1\frac{3}{8}$ inch in diameter, and is the same as the larger. It has been given to field officers, or officers of equal rank, and to such others as had succeeded to the actual command of a battalion, or of a corps equivalent to a battalion, during an engagement, in consequence of the death or removal of the original commander, and is worn from a similar riband from the button-hole of their uniform."

In consequence of many officers having received several medals, it became inconvenient to wear them.

The following regulations on the subject were issued :—

London Gazette, 9th October, 1813.

" Whereas considerable inconvenience has arisen from the number of medals which have been issued in commemoration of the brilliant and distinguished events in which the success of His Majesty's Arms has received the Royal approbation, the Prince Regent has been pleased to command that the following regulations shall be adopted in the grant and circulation of such marks of distinction, viz. :—

" 1st.—That one medal only shall be borne by each officer recommended for such distinction.

" 2nd.—That for the second and third events, which may be subsequently commemorated in like manner, each individual recommended to bear the distinction shall carry a gold clasp attached to the riband to which the medal is suspended, and inscribed with the name of the battle or siege to which it relates.

" 3rd.—That upon a claim being admitted to a fourth mark of distinction, a *cross* shall be borne by each officer, with the names of the four battles or sieges respectively inscribed thereupon ; and to be worn in substitution of the distinctions previously granted to such individuals.

" 4th.—Upon each occasion of a similar nature that may occur, subsequently to the grant of a cross, the clasp shall again be issued to those who have a claim to the additional distinction, to be borne on the riband to which the cross is suspended, in the same manner as described in No. 2 of these regulations.

" His Royal Highness was further pleased to command that the distribution of medals or badges for military services of distinguished merit shall be regulated as follows, viz. :—

" 1st.—That no general or other officer shall be considered entitled to receive them, unless he has been personally and particularly engaged upon those occasions of great importance and peculiar brilliancy, in commemoration of which, such marks of distinction are to be bestowed.

"2nd.—That no officer shall be considered a candidate for the medal or badge, except upon the special selection and report of the commander of the forces upon the spot, as having merited the distinction by conspicuous services.

"3rd.—That the commander of the forces shall transmit to the Commander-in-Chief returns, signed by himself, specifying the name and ranks of those officers whom he shall have selected as particularly deserving.

"4th.—The commander of the forces in making selection will restrict his choice to the undermentioned ranks, viz. :—

> "General officers.
>
> "Commanding officers of brigades.
>
> "Commanding officers of artillery or engineers.
>
> "Adjutant-general and quarter-master-general.
>
> "Deputies of adjutant-general and quarter-master-general having the rank of field officers.
>
> "Assistants, adjutant and quarter-master-generals, having the rank of field officers, and being at the head of the staff with a detached corps or distinct division of the army.
>
> "Military secretary having the rank of field officer.
>
> "Commanding officers of battalions or corps equivalent thereto, and officers who may have succeeded to the actual command during the engagement, in consequence of the death or removal of the original commanding officer.

"The crosses, medals, and clasps are to be worn by the general officers, suspended by a riband of the colour of the sash, with a blue edge, round the neck; and by the commanding officers of battalions, or corps equivalent thereto, and officers who may have succeeded to the actual command during the engagement; the chiefs of military departments, and their deputies and assistants (having the rank of field officers), and such other officers as may be specially recommended, attached to a riband of the same description, to the button-hole of their uniform.

"Those badges which would have been conferred upon the

officers who had fallen at, or died since, the battles and sieges, shall, as a token of respect for their memories, be transmitted to their respective families."

For the actions for which these badges, etc., were bestowed, *vide* page 8; being the same for which Her Majesty subsequently granted the war medal, with the exception of the gold medal given for the battle of Maida, which was entirely different.

MAIDA

Obverse.—Laureated head of His Majesty " Georgius tertius."

Reverse.—Britannia brandishing a spear in her right hand; in her left, a shield with the crosses of the Union, a figure of Victory hovering over her and crowning her with a laurel wreath; behind her a triquetra (emblem of Sicily); before her, " Maida, IVL. IV., MDCCCVI."

Only seventeen Officers received this Medal.

GOLD CROSS.

1.—Colonel John Keane.

On the Cross
- " Martinique."
- " Vittoria."
- " Orthes."
- " Toulouse."

Clasps ...
- " Pyrenees."
- " Nivelle."

2.—Lieutenant-Colonel Thos. Arbuthnot, Assistant Quarter-Master-General.

On the Cross
- " Roleia. Vimiera.
- " Corunna."
- " Pyrenees."
- " Nivelle."

Clasps ...
- " Corunna."
- " Orthes."

GOLD MEDAL.

(SMALL SIZE.)

JAVA.

3.—Capt. G. H. Gall, commanding Governor-General's-body-guard.

WAR MEDALS, 1793 TO 1814.

General Order. Horse Guards,
1st June, 1847.

" Her Majesty having been graciously pleased to command that a medal should be struck to record the services of her fleets and armies, during the wars commencing 1793, and ending in 1814, and that one should be conferred upon every officer, non commissioned officer, and soldier, of the army, who was in any battle or siege, to commemorate which, medals have been struck by command of Her Majesty's royal predecessors, and had been distributed to the general or superior officers of the several armies and corps of troops engaged, in conformity with the regulations of the army at that time in force, etc."

The following are the actions and sieges for which clasps were given :—

Maida, July 4th, 1806.

Roleia, 17th August, 1808.

Vimiera, 21st August, 1808.

Sahagun, December, 1808.

Benevente, January, 1809.

Corunna, 16th January, 1809.

*Martinique, February, 1809.

Talavera, 27th, 28th July, 1809.

*Guadaloupe, January, February, 1810.

Busaco, 27th September, 1810.

Barrosa, 5th March, 1811.

Fuentes D'Onor, 5th May, 1811.

Albuhera, 11th May, 1811.

Java, August, September 11th (taken from the Dutch.)

Cindad Rodrigo, January, 1812.

Badajoz, 17th March to 6th April, 1812.

Salamanca, 22nd July, 1812.

†Fort Detroit, August, 1812.

†Chateaugay, 26th October, 1813.

†Chrystler's Farm, 11th November, 1813.

Vittoria, 21st June, 1813.

Pyrenees, 28th July to 2nd August, 1813.

St. Sebastian, August & September.

Nivelle, 16th November.

Nive, 9th to 13th December.

Orthes, 27th February, 1814.

Toulouse, 10th April, 1814.

---

* French West India Island.   † North America.

EGYPT, 1801.

The clasp for the war in Egypt was not granted till 12th February, 1850, " To those who were still alive."

Obverse.—" Victoria Regina." Head of the Queen crowned; 1848 in the exergue.

Reverse.—" To the British Army." The Queen placing a laurel wreath upon the head of the Duke of Wellington, who kneels before her ; 1793–1814 in the exergue.

Riband : Crimson, with blue border.

One Clasp.

EGYPT.

4.—William Crozier	...	12th Light Dragoons.
5.—G. Midcalf	... ...	10th Foot.
6.—N. Gillespie	... ...	13th Foot.
7.—J. Rogers	... ...	80th Foot.
8.—J. Dennis	... ...	90th Foot.
9.—J. McNeil	... ...	92nd Foot.
10.—J. Swanwick	... ...	Queen's German Regiment.

Campaign in Egypt, 1801; under Sir R. Abercromby, K.B. The French commanded by General Abdallah Jaques Menou.

MAIDA.

11.—James Ellis	...	...	... 35th Foot.
12.—Thos. Marke	...	...	... 81st Foot.

Battle fought in Calabria, 4th July, 1806; under Major-General John Stuart. The French commanded by General Regnier.

ROLEIA.

13.—James Bennett ...	...	...	91st Foot.

Battle fought in Portugal, 17th August, 1808; under Lieut.-General The Hon. Sir Arthur Wellesley, K.B. The French commanded by General Laborde.

VIMIERA.

14.—JOHN BUDGELL ... ... 9th Foot.

Battle in Portugal, 21st August, 1808, under Lieutenant-General the Honourable Arthur Wellesley, K.G. The French commanded by Marshal Junot.

SAHAGUN AND BENEVENTE.*

15.—J. DRUMMOND ... ... 18th Light Dragoons.

Cavalry actions in Spain: " Sahagun," 20th December, 1808, under Lieutenant-General Lord Paget. The French commanded by Brigadier-General Debelle. " Benevente," 29th December, 1808, under Lieutenant-General Lord Paget and Brigadier-General the Honourable Charles Stewart. The French commanded by General Lefevre Desnouettes.

15A.—" Sahagun.'

CORUNNA.

16.—JOSEPH SIMMONS ... ... 14th Foot.
17.—E. KENNY, Sergeant ... ... 26th Foot.
CASE XII., No. 13.

Battle in Spain, 16th January, 1809 ; under Lieutenant-General Sir John Moore, K.B. The French commanded by Marshal Soult.

---

* There is a clasp inscribed " Sahagun " alone, given to those who were present on that occasion, and not at Benevente; but there was no clasp issued for " Benevente " alone.

MARTINIQUE.

18.—Daniel Banfield, Serjeant ... 8th Foot.
19.—Thomas Walsh ... ... 46th Foot.
20.—Sir C. F. Smith ... ... Royal Engineers.

Case XII., No. 34.

French West India Island; surrendered 24th February, 1809, to Lieutenant-General Beckwith. The French under Admiral Villaret-Joyeuse.

TALAVERA.

21.—J. Nicholls ... ... 23rd Light Dragoons.
Battle in Spain, 27th, 28th July, 1809; under Lieutenant-General Sir A. Wellesley. French commanded by King Joseph Bonaparte.

GUADALOUPE.

22.—A. Sturton ... ... 15th Foot.
23.—D. Evans, Corporal ... ... 96th Foot.
French West India Island; surrendered to Lieutenant-General Sir George Beckwith, 6th February, 1810. The French commanded by General Ernoufe.

BUSACO.

24.—John McGiffard ... ... 74th Foot.
Battle in Portugal, 27th September, 1810; under Lieutenant-General Viscount Wellington. The French commanded by Marshal Massena.

BARROSA.

25.—John Moore ... ... 67th Foot.
Battle in Spain, 5th March, 1811; under Lieutenant-General Thomas Graham. The French under Marshal Victor.

FUENTES D'ONOR.

26.—David Farrell ... ... 85th Foot.
Battle in Spain, 5th May, 1811; under Lieutenant-General Viscount Wellington. The French under Marshal Massena.

ALBUHERA.

27.—J. Reid ... ... 29th Foot.
Battle in Spain, 16th May, 1811; under Major-General Sir William Beresford. The French under Marshal Soult.

JAVA.

28.—Thomas Smith ... ... 59th Foot.
29.—D. McLeod, Sergeant ... ... 78th Foot.
Case XII., No. 34.

30.—Patrick Carroll ... ... 89th Foot.
Taken 18th September, 1811, by Lieutenant-General Sir Samuel Auchmuty, from the Dutch, commanded by General Jansens.

CIUDAD RODRIGO.

31.—John Kelly ... ... 7th Foot.
Fortress in Spain. Taken 19th January, 1812, by Lieutenant-General Viscount Wellington. The French under General Barrié.

BADAJOZ.

32.—D. Keegan ... ... 30th Foot.
33.—T. Welsh, Sergeant ... ... 77th Foot.

Spanish Fortress. Invested 17th March; taken by storm, 6th April, 1812, by Lieutenant-General Earl of Wellington. French, under General Phillippon, surrendered 7th April.

SALAMANCA.

34.—T. Jones ... ... 1st Foot, Royals.
35.—T. Bradshaw, Troop Sergt.-Maj... 11th Light Dragoons.
Case XII., No. 4.

Battle in Spain, 22nd July, 1812, under General, the Earl of Wellington. French under Marshal Marmont, Duke of Ragusa.

FORT DETROIT.

36.—R. James ... ... 41st Foot.

A fort on the St. Lawrence, America. Surrendered to Major-General Brock, 16th August, 1812. The Americans commanded by General Hull.

VITTORIA.

37.—Patrick McGowan .. ... 68th Foot.

Battle in Spain, 21st June, 1813; under General the Marquis of Wellington. The French under King Joseph Bonaparte, and Marshal Jourdan.

PYRENEES.

38.—ALEXANDER McINTOSH ...   ...   71st Foot.

July 28th to 2nd August, 1813, on the French and Spanish frontier ; under Field-Marshal the Marquis of Wellington. The French under Marshal Soult.

ST. SEBASTIAN.

39.—J. BUTTERWORTH   ...   ...   1st Foot Guards.

Spanish fortress. Surrendered 9th September, 1813, to Field-Marshal the Marquis of Wellington. The French under General Rey.

CHATEAUGAY.

40.—SAKKARIWAKERON   ... - ...   Warrior.

Battle on the Chateaugay River, Lower Canada, 26th October, 1813 ; under Lieutenant-General De Saluberry. The Americans under General Hampton.

NIVELLE.

41.—Lieutenant W. CRAWLEY   ...   27th Foot.

Battle in Spain, 10th November, 1813 ; under Field-Marshal the Marquis of Wellington. The French under Marshal Soult.

CHRYSTLER'S FARM.

42.—James Rooke   ...   . 49th Foot.

Battle near Chrystler's Farm, Upper Canada, 11th November, 1813; under Lieutenant-Colonel J. L. Morrison, 89th Foot. The Americans under Major-General Boyd.

NIVE.

43.—Geo. Wilcock ...   . 84th Foot.

9th to 13th December, 1813, on the French frontier; under Lieutenant-General Sir Rowland Hill. The French under Marshal Soult.

ORTHES.

44.—Joseph Heath ...   . 23rd Foot.

Battle in France, 27th February, 1814; under Field-Marshal the Marquis of Wellington. The French under Marshal Soult.

TOULOUSE.

45.—Wilson Harrison   ..   ... 1st Royal Dragoons.
46.—E. Holland, Clerk   ...   ... P. M. G. Dep.

The final battle of the war, fought in France 10th April, 1814; under Field-Marshal the Marquis of Wellington. The French under Marshal Soult.

Two Clasps.

VITTORIA, TOULOUSE.

47.—John Driver	1st Life Guards.
48.—J. Stanfield	2nd Life Guards.
49.—James Hayes	Royal Horse Guards.
50.—R. Forknall	3rd Dragoon Guards.
51.—John Clarke	3rd Light Dragoons.

FUENTES D'ONOR, VITTORIA.

52.—Robert Waller	14th Light Dragoons.

SALAMANCA, VITTORIA.

53.—T. Armstrong, Corporal ...	16th Light Dragoons.

ROLEIA, VIMIERA.

54.—Thomas Druce	20th Light Dragoons.

NIVELLE, NIVE.

55.—T. Griffiths	1st Foot Guards.
56.—E. King	Coldstream Guards.
57.—J. Cooms	62nd Foot.

CIUDAD RODRIGO, SALAMANCA.

58.—J. Lynn ...   . 2nd Foot Guards.

EGYPT, MAIDA.

59.—Owen Reilly ...   . 27th Foot.

BUSACO, ALBUHERA.

60.—William Regan ...   . 29th Foot.

VITTORIA, ST. SEBASTIAN.

61.—W. Hogg, Corporal   ...   ... Royal H. Artillery.
Case XII., No. 32.

EGYPT, TALAVERA.

62.—Anthony Allen ...   . 61st Foot.

MAIDA, JAVA.

63.—John Scott   ...   . 78th Foot.

TALAVERA, TOULOUSE.

64.—Martin Durkin ...   . 87th Foot.

ORTHES, TOULOUSE.

65.—James Weedy   ...   . 10th Hussars.
66.—Edward Lynch ...   . Hospital Mate.

MARTINIQUE, GUADALOUPE.

67.—William Skerrett   ...   ... 25th Foot.

Three Clasps.

TALAVERA, ALBUHERA, TOULOUSE.

68.—JOHN BARKER ... ... 3rd Dragoon Guards.

SALAMANCA, VITTORIA, TOULOUSE.

69.—J. McALISTER ... ... 5th Dragoon Guards.

SAHAGUN AND BENEVENTE, VITTORIA, TOULOUSE.

70.—THOMAS THOMPSON ... ... 10th Hussars.

ALBUHERA, VITTORIA, TOULOUSE.

71.—GEORGE HOLMES ... ... 13th Light Dragoons.

CORUNNA, NIVELLE, NIVE.

72.—F. DAWSON ... ... 1st Foot Guards.
73.—THOS. WHITE ... ... 76th Foot.

VITTORIA, ST. SEBASTIAN, NIVELLE.

74.—Robert Reid ... ... 1st Foot.

SALAMANCA, VITTORIA, ST. SEBASTIAN.

75.—R. Munday ... ... 2nd Foot.

BADAJOZ, VITTORIA, ST. SEBASTIAN.

76.—James Callyer ... ... 4th Foot.

ROLEIA, VIMIERA, CORUNNA.

77.—John Brown ... ... 6th Foot.
78.—David Dewar ... ... 91st Foot.
Case XII., No. 27.

VITTORIA, ORTHES, TOULOUSE.

79.—Edward Cooper ... ... 20th Foot.

EGYPT, MARTINIQUE, GUADALOUPE.

80.—Thomas Crotty ... ... 25th Foot.

BARROSA, VITTORIA, TOULOUSE.

81.—Stephen Hill ... ... 28th Foot.

VIMIERA, SALAMANCA, PYRENEES.

82.—William Fox ... ... 36th Foot.

SALAMANCA, VITTORIA, ST. SEBASTIAN.

83.—Richard Cross ... ... 38th Foot.

VIMIERA, CORUNNA, SALAMANCA.

84.—James Hancock ... ... 36th Foot.

VITTORIA, PYRENEES, TOULOUSE.

85.—James Thompson ... ... 39th Foot.

ALBUHERA, VITTORIA, TOULOUSE.

86.—M. Farraker ... ... 57th Foot.

VITTORIA, ST. SEBASTIAN, NIVE.

87.—Joshua Lord ... ... 59th Foot.
Case XII., No. 20.

VIMIERA, CORUNNA, FUENTES D'ONOR.

88.—James Kennan ... ... 71st Foot.

CORUNNA, BUSACO, FUENTES D'ONOR.

89.—JOHN GAYLOR ... ... 79th Foot.

NIVELLE, NIVE, TOULOUSE.

90.—GEORGE McKAY ... ... 79th Foot.
CASE XII., No. 24B.

SALAMANCA, VITTORIA, PYRENEES.

91.—J. HACKETT ... ... 95th Foot.
CASE XII., No. 29.

CIUDAD RODRIGO, BADAJOZ, SALAMANCA.

92.—J. LAITEY ... ... Royal Staff Corps.

FUENTES D'ONOR, VITTORIA, TOULOUSE.

93.—T. DUEN ... ... 1st Dragoons.
CASE XII., No. 1.

EGYPT, CIUDAD RODRIGO, BADAJOZ.

94.—J. MACKEY... ... ... Rl. Saps. and Miners.

Four Clasps.

SAHAGUN AND BENEVENTE, NIVE, ORTHES, TOULOUSE.

95.—T. Seaman ... ... 7th Hussars.

SAHAGUN, VITTORIA, ORTHES, TOULOUSE.

96.—R. R. Handford ... ... 15th Hussars.

SAHAGUN AND BENEVENTE, VITTORIA, ORTHES, TOULOUSE.

97.—K. Jameson ... ... 18th Hussars.

BARROSA, CIUDAD RODRIGO, SALAMANCA, VITTORIA.

98.—W. Westerman ... ... 3rd Foot Guards.

VITTORIA, PYRENEES, NIVELLE, ORTHES.

99.—William Johnston ... ... 6th Foot.

ROLEIA, VIMIERA, TALAVERA, ALBUHERA.

100.—Thomas Robson ... ... 29th Foot.

Case XII., No. 43.

ROLEIA, VIMIERA, CORUNNA, SALAMANCA.

101.—William Parsley ... ... 32nd Foot.

VITTORIA, ST. SEBASTIAN, NIVELLE, NIVE.

102.—H. Sharrock ... ... 47th Foot.

PYRENEES, NIVELLE, ORTHES, TOULOUSE.

103.—S. Drought ... ... 48th Foot.

FUENTES D'ONOR, ST. SEBASTIAN, NIVELLE, NIVE.

104.—Edward McLear ... ... 85th Foot.

CORUNNA, FUENTES D'ONOR, VITTORIA, PYRENEES.

105.—J. McLean ... ... 92nd Foot.

SALAMANCA, VITTORIA, PYRENEES, TOULOUSE.

———

106.—N. HASLEM ... ... Gunner Royal H.A.

———

ALBUHERA, SALAMANCA, VITTORIA, ST. SEBASTIAN.

———

107.—ECKHARD BOHNE ... ... 2nd Light Bat. K.G.L.

———

CORUNNA, BUSACO, FUENTES D'ONOR, SALAMANCA.

———

108.—H. BANNERMAN, Sergeant ... 79th Foot.
CASE XII., No. 24.

Five Clasps.

TALAVERA, ALBUHERA, SALAMANCA, VITTORIA, TOULOUSE.

109.—T. Moore   ...   .   4th Light Dragoons.

TALAVERA, BUSACO, ALBUHERA, PYRENEES, TOULOUSE.

110.—Charles Maycock   ...   ...   3rd Foot.

CORUNNA, SALAMANCA, PYRENEES, ORTHES, TOULOUSE.

111.—J. McLean   ...   .   42nd Foot.*

SALAMANCA, PYRENEES, NIVELLE, NIVE, ORTHES.

112.—W. Murphy, Sergeant   ...   42nd Foot.
Case XII., No. 16.

---

* This was one of the men who carried Sir John Moore to the grave.

VIMIERA, TALAVERA, CIUDAD RODRIGO, VITTORIA, PYRENEES.

113.—Thos. Clements, Sergeant ... 40th Foot.

SALAMANCA, VITTORIA, PYRENEES, ORTHES, TOULOUSE.

114.—C. Bishop, Drummer ... 40th Foot.

SALAMANCA, VITTORIA, PYRENEES, ST. SEBASTIAN, ORTHES.

115.—T. Quinlan ... ... 68th Foot.

FUENTES D'ONOR, VITTORIA, PYRENEES, NIVE, ORTHES.

116.—Alexander Graham ... ... 71st Foot.
Case XII., No. 21.

PYRENEES, NIVELLE, NIVE, ORTHES, TOULOUSE.

117.—Walter Ross... ... ... 91st Foot.

CORUNNA, VITTORIA, PYRENEES, NIVE, ORTHES.

118.—M. NICOLSON...   ...   ... 92nd Foot.*

ALBUHERA, VITTORIA, PYRENEES, ORTHES, TOULOUSE.

119.—J. ASHMORE   ...   . Royal H. Artillery.

---

* Private Nicolson lost his arm at Orthes, died in the Isle of Skye in 1863, aged 90. He went by the name of " Calum Na Ritaig," or Malcolm with the stump arm.

Six Clasps.

BUSACO, FUENTES D'ONOR, CIUDAD RODRIGO, SALAMANCA, VITTORIA, ST. SEBASTIAN.

120.—J. Lovett ... ... Coldstream Guards.

TALAVERA, BUSACO, CIUDAD RODRIGO, SALAMANCA, VITTORIA, NIVELLE.

121.—James Malther ... ... Coldstream Guards.

TALAVERA, BUSACO, FUENTES D'ONOR, CIU-DAD RODRIGO, SALAMANCA, VITTORIA.

122.—J. Drysdale ... ... 3rd Foot Guards.

BUSACO, PYRENEES, NIVELLE, NIVE, ORTHES, TOULOUSE.

123.—Alexander Fields ... ... 11th Foot.

ALBUHERA, BADAJOZ, SALAMANCA, VITTORIA, PYRENEES, TOULOUSE.

124.—George Hoole   ...   ... 23rd Foot.

ALBUHERA, VITTORIA, PYRENEES, NIVELLE, NIVE, ORTHES.

125.—Joseph Dodd ...   . 34th Foot.

TALAVERA, BUSACO, ALBUHERA, CIUDAD RODRIGO, BADAJOZ, SALAMANCA.

126.—W. Clayton, Sergeant   ... 48th Foot.

SALAMANCA, VITTORIA, PYRENEES, ST. SEBASTIAN, NIVELLE, ORTHES.

127.—John Galvin ...   . 51st Foot.

FUENTES D'ONOR, SALAMANCA, PYRENEES, ST. SEBASTIAN, NIVELLE, ORTHES.

128.—J. Greenwood, Sergeant   ... 51st Foot.

Case XII., No. 19.

CIUDAD RODRIGO, SALAMANCA, VITTORIA, PYRENEES, ORTHES, TOULOUSE.

129.—T. INGHAM ... ... 52nd Foot.

TALAVERA, BUSACO, ALBUHERA, VITTORIA, PYRENEES, TOULOUSE.

130.—MOSES BUTTON ... ... 66th Foot.

CIUDAD RODRIGO, SALAMANCA, VITTORIA, PYRENEES, ORTHES, TOULOUSE.

131.—T. GIBBS, Gunner ... ... Royal H. Artillery.

CORUNNA, BADAJOZ, VITTORIA, PYRENEES, ORTHES, TOULOUSE.

132.—T. HANDCOCKS, Sergeant ... Royal Waggon Train.

Seven Clasps.

TALAVERA, BUSACO, FUENTES D'ONOR, CIUDAD RODRIGO, SALAMANCA, VITTORIA, NIVE.

133.—C. RAWLINS ... ... 3rd Foot Guards.

ROLEIA, VIMIERA, CORUNNA, BUSACO, FUENTES D'ONOR, SALAMANCA, VITTORIA.

134.—CHAS. LACEY ... ... 9th Foot.

BARROSA, VITTORIA, PYRENEES, NIVELLE, NIVE, ORTHES, TOULOUSE.

135.—WILLIAM MOORE ... ... 87th Foot.

TALAVERA, BUSACO, SALAMANCA, PYRENEES, NIVE, ORTHES, TOULOUSE.

136.—E. JENNINGS, Sergeant ... 61st Foot.

ROLEIA, VIMIERA, CORUNNA, VITTORIA, PYRENEES, NIVELLE, ORTHES.

—

137.—E. LEATHER, Sergeant   ... 82nd Foot.

—

FUENTES D'ONOR, CIUDAD RODRIGO, BADAJOZ, SALAMANCA, VITTORIA, PYRENEES, TOULOUSE.

—

138.—WILLIAM GRANT   ...   ... 94th Foot.

—

Eight Clasps.

TALAVERA, BUSACO, BADAJOZ, SALAMANCA, VITTORIA, PYRENEES, ORTHES, TOULOUSE.

139.—ROBERT BOWKER ... ... 13th Light Dragoons.

TALAVERA, BUSACO, FUENTES D'ONOR, CIUDAD RODRIGO, SALAMANCA, VITTORIA, NIVELLE, NIVE.

140.—CHARLES WESTON, Qrmaster.... 3rd Foot Guards.

VIMIERA, CORUNNA, SALAMANCA, VITTORIA, PYRENEES, NIVELLE, NIVE, TOULOUSE.

141.—E. LAWRENCE, Drummer ... 5th Foot.

TALAVERA, BUSACO, FUENTES D'ONOR, CIUDAD RODRIGO, SALAMANCA, VITTORIA, PYRENEES, ORTHES.

142.—JOHN ARNOLD ... ... 24th Foot.

ALBUHERA, BADAJOZ, SALAMANCA, VITTORIA,
PYRENEES, NIVELLE, ORTHES, TOULOUSE.

———

143.—ALEXANDER FRASER, Drummer   27th Foot.

———

TALAVERA, BUSACO, ALBUHERA, VITTORIA,
PYRENEES, NIVE, ORTHES, TOULOUSE.

———

144.—G. PRICKETT ...   .   31st Foot.

———

CORUNNA,   CIUDAD   RODRIGO,   BADAJOZ,
SALAMANCA,   VITTORIA,   PYRENEES,   ST.
SEBASTIAN, ORTHES.

———

145.—GEORGE FROST   ...   ...   52nd Foot.

———

FUENTES D'ONOR, CIUDAD RODRIGO, BADAJOZ,
VITTORIA,   PYRENEES,   NIVELLE,   NIVE,
ORTHES.

———

146.—WILLIAM CLARKE   ...   ...   94th Foot.

———

BARROSA, CIUDAD RODRIGO, BADAJOZ, SALA-
MANCA,   VITTORIA,   PYRENEES,   ORTHES,
TOULOUSE.

———

147.—W. BURROWS ...   .   95th Foot.

Nine Clasps.

TALAVERA, BUSACO, BARROSA, FUENTES
D'ONOR, CIUDAD RODRIGO, SALAMANCA,
VITTORIA, NIVELLE, NIVE.

148.—J. Biddle ... ... 3rd Foot Guards.

MARTINIQUE, BUSACO, ALBUHERA, CIUDAD
RODRIGO, BADAJOZ, VITTORIA, PYRENEES,
ORTHES, TOULOUSE.

149.—J. Whitworth, Sergeant ... 7th Foot.

EGYPT, CORUNNA, FUENTES D'ONOR, SALA-
MANCA, PYRENEES, NIVELLE, NIVE,
ORTHES, TOULOUSE.

150.—J. Mackenzie, Corporal ... 42nd Foot.

TALAVERA, BUSACO, ALBUHERA, CIUDAD
RODRIGO, BADAJOZ, SALAMANCA, VITTORIA,
PYRENEES, TOULOUSE.

151.—W. Clark ... ... 48th Foot.

BUSACO, FUENTES D'ONOR, CIUDAD RODRIGO, BADAJOZ, SALAMANCA, PYRENEES, NIVE, ORTHES, TOULOUSE.

152.—John Stanage...   ...   ... 52nd Foot.

FUENTES D'ONOR, CIUDAD RODRIGO, BADAJOZ, SALAMANCA, VITTORIA, PYRENEES, NIVE, ORTHES, TOULOUSE.

153.—Brian Mullony   ...   ... 52nd Foot.

TALAVERA, BUSACO, FUENTES D'ONOR, CIUDAD RODRIGO, SALAMANCA, VITTORIA, NIVELLE, ORTHES, TOULOUSE.

154.—James Rigney...   ...   ... 83rd Foot.

FUENTES D'ONOR, BADAJOZ, SALAMANCA, VITTORIA, PYRENEES, NIVELLE, NIVE, ORTHES, TOULOUSE.

155.—James Edmunds   ...   ... 94th Foot.

Ten Clasps.

BUSACO, ALBUHERA, CIUDAD RODRIGO, BADAJOZ, SALAMANCA, VITTORIA, ST. SEBASTIAN, NIVELLE, ORTHES, TOULOUSE.

156.—J. BOYDALL ... ... R. Artillery Drivers.

VIMIERA, BUSACO, FUENTES D'ONOR, CIUDAD RODRIGO, BADAJOZ, SALAMANCA, VITTORIA, PYRENEES, ORTHES, TOULOUSE.

157.—MARTIN MURRAY ... ... 52nd Foot.

BUSACO, FUENTES D'ONOR, CIUDAD RODRIGO, BADAJOZ, SALAMANCA, VITTORIA, PYRENEES, NIVELLE, ORTHES, TOULOUSE.

158.—D. McGURK, Sergeant ... 74th Foot.*

* This man was wounded at the Battle of Vittoria. Rec. 74th High. Page 137.

Eleven Clasps.

TALAVERA, BUSACO, FUENTES D'ONOR, BADAJOZ, SALAMANCA, VITTORIA, PYRENEES, NIVELLE, NIVE, ORTHES, TOULOUSE.

159.—WILLIAM HANLEY, Troop Sergt.-Major ... ... 14th Light Dragoons.
CASE XII., No. 5.

CORUNNA, BUSACO, FUENTES D'ONOR, CIUDAD RODRIGO, BADAJOZ, SALAMANCA, VITTORIA, NIVELLE, NIVE, ORTHES, TOULOUSE.

160.—JOSEPH GREEN ... ... 52nd Foot.
CASE XII., No. 18.

BUSACO, FUENTES D'ONOR, CIUDAD RODRIGO, BADAJOZ, SALAMANCA, VITTORIA, PYRENEES, NIVELLE, NIVE, ORTHES, TOULOUSE.

161.—W. GRAHAM, Sergeant ... 74th Foot*

---

* Sergeant W. Graham was wounded at Toulouse. Hist. Records, 74th High. Page 139.

TALAVERA, BUSACO, FUENTES D'ONOR,
BADAJOZ, SALAMANCA, VITTORIA,
PYRENEES, NIVELLE, NIVE, ORTHES,
TOULOUSE.

162.—JAMES TRACEY...   ...   ... 88th Foot.

CASE XII., No. 26.

ROLEIA, VIMIERA, BUSACO, CIUDAD RODRIGO,
BADAJOZ, SALAMANCA, PYRENEES,
ST. SEBASTIAN, NIVELLE, ORTHES,
TOULOUSE.

163.—T. SMITH   ...   . 95th Foot, Rifles.

BUSACO, FUENTES D'ONOR, CIUDAD RODRIGO,
BADAJOZ, SALAMANCA, VITTORIA,
PYRENEES, NIVELLE, NIVE, ORTHES,
TOULOUSE.

164.—H. BOOTH   ...   . 95th Rifles.

Twelve Clasps.

ROLEIA, VIMIERA, TALAVERA, BUSACO, CIUDAD RODRIGO, DADAJOZ, SALAMANCA, VITTORIA, PYRENEES, NIVELLE, ORTHES, TOULOUSE.

165.—James Hogan ...　...　... 45th Foot.

CORUNNA, BUSACO, FUENTES D'ONOR, CIUDAD RODRIGO, BADAJOZ, SALAMANCA, VITTORIA, PYRENEES, NIVELLE, NIVE, ORTHES, TOULOUSE.

166.—B. Housely, Sergt. ...　... 52nd Foot.
Case XII., No. 17.

VIMIERA, BUSACO, FUENTES D'ONOR, CIUDAD RODRIGO, BADAJOZ, SALAMANCA, VITTORIA, PYRENEES, NIVELLE, NIVE, ORTHES, TOULOUSE.

167.—John Porter ...　...　... 52nd Foot.

TALAVERA, BUSACO, FUENTES D'ONOR, CIUDAD RODRIGO, BADAJOZ, SALAMANCA, VITTORIA, PYRENEES, NIVELLE, NIVE, ORTHES, TOULOUSE.

168.—J. HOLLOWAY ...　...　... 88th Foot.

Thirteen Clasps.

MARTINIQUE, BUSACO, ALBUHERA, CIUDAD
RODRIGO, BADAJOZ, SALAMANCA, VITTORIA,
PYRENEES, ST. SEBASTIAN, NIVELLE
NIVE, ORTHES, TOULOUSE.

169.—Peter Hardy, Corporal   ...   7th Foot.*
170.—James Hardy ...   .   7th Foot.*

VIMIERA, BUSACO, FUENTES D'ONOR,
CIUDAD RODRIGO, BADAJOZ, SALAMANCA,
VITTORIA, PYRENEES, ST. SEBASTIAN,
NIVELLE, NIVE, ORTHES, TOULOUSE.

171.—J. Faloon   ...   .   52nd Foot.

---

* These two men were brothers of Sergeant J. Hardy, No. 172, and all served
together in the same Regiment.

Fourteen Clasps.

MARTINIQUE, TALAVERA, BUSACO, ALBU-
HERA, CIUDAD RODRIGO, BADAJOZ,
SALAMANCA, VITTORIA, PYRENEES, ST.
SEBASTIAN, NIVELLE, NIVE, ORTHES,
TOULOUSE.

172.—J. Hardy, Sergèant ...   ...   7th Foot.*

ROLEIA, VIMIERA, TALAVERA, BUSACO,
FUENTES D'ONOR, CIUDAD RODRIGO,
BADAJOZ, SALAMANCA, VITTORIA,
PYRENEES, NIVELLE, NIVE, ORTHES,
TOULOUSE.

173.—J. Nixon, Sergeant ...   ...   48rd Foot.

* Sergeant Hardy joined the 7th Fusiliers in 1808 ; was present at Copenhagen,
1807 ; at the storming of Badajoz he distinguished himself by carrying Lieutenant-
Col. Spottiswoode, 71st Light Infantry, Aide-de-Camp to Major General Colville,
out of the breach where he lay severely wounded.   For this act he received a
pension for life from the family of Spottiswoode.   He was a volunteer, at the
storming of St. Sebastian, to the column of assault.   In 1814, he proceeded with
his regiment to America ; was engaged in the attack on New Orleans ; served the
campaign of 1815 ; was present at Waterloo, having accompanied an officer of
the 7th Fusiliers on staff employ.   He died at the age of 94, in 1876.

Fifteen Clasps.

ROLEIA, VIMIERA, CORUNNA, TALAVERA, BUSACO, FUENTES D'ONOR, CIUDAD RODRIGO, BADAJOZ, SALAMANCA, VITTORIA, PYRENEES, NIVELLE, NIVE, ORTHES, TOULOUSE.

174.—JAMES TALBOT...   ...   ... 45th Foot.

CASE II.

WATERLOO.

Horse Guards,
10th March, 1816.

The Prince Regent has been graciously pleased, in the name and on behalf of His Majesty, to command that, in commemoration of the brilliant and decisive victory of Waterloo, a medal should be conferred on every officer, non-commissioned officer, and soldier of the British Army present upon that memorable occasion, etc.

This medal was given to every man present on three days, viz. :—

Quatre Bras	16th June, 1815.
Action on	17th ,,
Waterloo	18th ,,

Obverse.—Laureated head of the Prince Regent. "George P. Regent."

Reverse.—"Wellington" above a figure of Victory, seated with outspread wings—in her right hand a palm, in her left an olive branch. "Waterloo," with the date June 18th, 1815, in the exergue.

Riband: Crimson, with blue border.

WATERLOO.

CAVALRY.

1.—John Denby	1st Regiment of Life Guards.
2.—George Clayton, Farrier	2nd ,, ,,
3.—James Ogilway... ...	Royal Horse Guards.
4.—Charles Lord	1st Dragoon Guards.
5.—Joseph Keeling ...	1st Royal Dragoons.
6.—John Duen	,, ,,

Case XII., No. 1.

H

7.—JOHN TANNOCK, Sergeant   2nd R. N. British Dragoons.
8.—JOHN FOX, Sergeant   ...   6th, or Inniskilling   „
9.—JOHN MANN   ...   ...   7th Regiment Hussars.
10.—JOHN FREEMAN ...   ...   10th Royal Regiment Hussars.
11.—MATHEW BURMAN   ...   11th Regiment Light Dragoons.
12.—SAMUEL HALL ...   ...   12th   „   „
13.—JAMES COOPER ...   ...   13th   „   „
14.—JOHN SCOTT, Sergeant ...   15th King's Hussars.
15.—THOMAS HILTON...   ...   16th Queen's Light Dragoons.
16.—JAMES KELLY, 1st   ...   10th Regiment Hussars.
17.—THOMAS METCALFE   ...   23rd Regiment Light Dragoons.

ROYAL ARTILLERY.

18.—SAMUEL HALL, Corporal . Royal Foot Artillery.
19.—GEORGE BELL, Gunner   · „   „
20.—E. D. NEALE, Clerk of
    Stores   ...   ...   „   „
21.—WILLIAM FLEMING, Driver   Royal Artillery.
22.—D. HOGG, Corporal   ...   Royal Horse Artillery.
                                CASE XII., No. 32.
23.—JAMES MCKENZIE, Corpl.   „   „
24.—JAMES GIBB, Gunner   ...   „   „
25.—SILVESTER CLAIL, Driver   „   „
26.—EDWARD BROWN, Smith   „   „ .
27.—RICHARD BANT, Assistant-
    Commissary...   ...   Field Train Department.
28.—JAMES TAYLOR ...   ...   Royal Waggon Train.

INFANTRY.

29.—EDWARD SPICER   ...   Grenadier Guards, 2nd Batt.
30.—FRANCIS MAYON   ...   Grenadier Guards, 3rd Batt.
31.—JAMES STANDLEY, Sergt.   Coldstream Guards, 2nd Batt.
32.—JOHN NEWCOMBES   ...   3rd Regt. of Guards, 2nd Batt.
33.—JOHN FORDYCE ...   ...   1st   Foot,   or   Royal   Scots,
                                3rd Batt.
34.—WILLIAM JONES, Sergt.   4th Regt. of Foot, 1st Batt.
35.—THOMAS BARBER   ...   14th Regt. of Foot, 3rd Batt.

36.—John Morris, Corporal 23rd Regt. of Foot, R. W. F.
37.—James McCouvelle ... 27th Regt. of Foot, 1st Batt.
38.—R. P. Eason, Lieut.... 28th Regt. of Foot.
39.—James Asguith, Drmmr. 30th Regt. of Foot, 2nd Batt.
40.—John Rodgers, Sergeant 32nd Regt. of Foot.
41.—Josiah Smith....   ... 33rd Regt. of Foot.
42.—John Pearce, Sergeant *35th Regt. of Foot, 2nd Batt.
43.—Joseph Edwards   ... 40th Regt. of Foot, 1st Batt.
44.—W. Murphy, Sergeant 42nd R. H. Regt. Infantry.
                                    Case XII., No. 16.

45.—Dougal McPherson ...   ,,   ,,   ,,
46.—Francis Dwyer   ... 44th Regt. of Foot, 2nd Batt.
47.—James McNeill   ... 51st Regt., Light Infantry.
48.—T. Greenwood, Sergt.   ,,   ,, Case XII., No. 19.

49.—   ... 52nd Regt. of Foot, 1st Batt.
50.—B. Housely, Sergeant   ,,   ,, Case XII., No. 17.

51.—Joseph Green   ...   ,,   ,, Case XII., No. 18.

52.—John Masters   ...   ,,   ,,   ,,
53.—J. H. Potts, Lieut. ... *54th Regt. of Foot, 1st Batt.
54.—Joshua Lord ...   ... *59th Regt. of Foot, 2nd Batt.
                                    Case XII., No. 20.

55.—Samuel Crawford   ...   ,,   ,,   ,,
56.—Richard Kew, Sergeant 69th Regt. of Foot, 2nd Batt.
57.—James Blair ...   ... 71st Regt. of Foot, 1st Batt.
58.—Alexander Graham   ...   ,,   ,, Case XII., No. 21.

59.—John Simmonds   ... 73rd Regt. of Foot, 2nd Batt.
60.—John Sutherland, Sergt. 79th Regt. of Foot, 1st Batt.
61.—Hugh Bannerman, Sergt.   ,,   ,, Case XII., No. 24.

62.—Jacobus Won-Der-Wer *91st Regt. of Foot, 1st Batt.
63.—David Dewar...   ...   ,,   ,, Case XII., No. 27.

---

* These four Regiments have not " Waterloo " on their colours, being the
regiments which formed Sir Charles Colville's Brigade, which was detached, but
the men received the medal.

64.—Donald Macdonald ...   92nd Highlanders, 1st Batt.
65.—Benjamin Slaughter...   95th Regt. of Foot, 1st Batt.
66.—John Rowland   ...   ,,   ,,   2nd ,,
67.—Robert Jackson   ...   ,,   ,,   3rd ,,
68.—J. Hackett ...   ...   ,,   ,,   ,, ,,

Case XII., No. 29.

THE KING'S GERMAN LEGION.

69.—Thomas Power, Vete-
     rinary Surgeon   ...   1st Regt. Hussars, K.G.L.
70.—Frederick Koch   ...   3rd Regt. Hussars, K.G.L.
71.—Conrad Oldershausen   1st Light Brigade, K.G.L.
72.—George Werner   ...   2nd Regt. Light Dragoons, K.G.L
73.—Henry Hillman   ...   1st Light Batt., K.G.L.
74.—Geo. Haasmann, Capt.   2nd Light Batt., K.G.L.
75.—Authur Apfelstadt,
     Drummer   ...   ...   1st Line Battalion, K.G.L.
76.—Fredk. Schmeck, Sergt.   2nd Line Batt., K.G.L.
77.—Frederick Kortencamp   3rd Line Batt., K.G.L.
78.—Lewis Engehausen, Cor.   4th Line Batt., K.G.L.
79.—Daniel Kamphuysen ...   5th Line Batt., K.G.L.
80.—Bernhard Schrage   ...   8th Line Batt., K.G.L.
81.—William Bohm   ...   King's German Artillery.

HANOVER. Waterloo Medal.

82.—Founded in 1817, by the Prince Regent of England; given to the soldiers of his hereditary dominions in Germany, who were present at Waterloo.

Obverse.—Head of Prince Regent. " George Prinz Regent," 1815.

Reverse.—Military trophy, "Waterloo, Jun. XVIII. Hanoverscher Tapferkeit."

Riband : Red, blue edge.

Tambour JOHAN MEYER,
Landwehr, Bataillon Hoya.

NASSAU.

83.—Founded by Duke Frederick, 23rd December, 1815.

Obverse.—Head of Duke " Friedrich August, Herzog zu Nassau."

Reverse.—Figure of Victory crowning a warrior. " Den Nassauischen Streitern bei Waterloo." Den 18 Jun. 1815.

Riband : Blue, yellow edge.

BRUNSWICK.

84.—Founded by the Prince Regent of Great Britain, as Guardian of the minor Princes of Brunswick, on the 11th June, 1818. " For the perpetual remembrance of the Campaign of 1815."

Obverse.—Head of Duke Frederick William (who fell at the battle of Quatre Bras, 16th June, 1815.)

Reverse.—1815, in a laurel wreath. " Braunschweig seinen Kriegern—Quatre Bras and Waterloo."

Riband : Yellow, light blue border.

Made out of the guns captured.

WILH. HANNE, Haute Corps.

CASE III.

EAST INDIA COMPANY'S MEDALS,
1784 TO 1857.

DECCAN, 1784.

1.—This medal was granted by an Order in Council, January 19th, 1794, to the troops under General Goddard; by a subsequent order, 22nd January, 1785, to the troops under Colonel Pearse, in the Carnatic.

Given in gold to subadars, silver to jemadars, and in inferior metal to lower ranks.

Obverse.—Britannia seated upon a trophy of arms, holding a wreath in her outstretched hand towards a fortress.

Reverse.—Persian inscription : " The courage and exertions of those valient men by whom the name of Englishmen has been celebrated and exhalted from Hindostan to the Deccan, having been established throughout the world, this has been granted by the Government of Calcutta, in commemoration of the excellent services of the brave. Year of the Hegira, 1199." A.D. 1784.

Within a circle is the following :—

" As coins are current in the world, so shall be the bravery and exploits of those heroes by whom the name of the victorious English nation was carried from Bengal to the Deccan."

Worn from a yellow cord.

MYSORE, 1791–1792.

Granted by an Order in Council, 4th June, 1793. The Company's troops commanded by Lord Cornwallis, General Medows, and General Abercromby. The enemy, by Tippoo Sultaun Bahadur.

This medal, like the preceding one, was given in gold, silver, etc.

Obverse.—A sepoy holding the British colours in his right hand ; in his left, the enemy's colours reversed ; in the background, a fortified city.

Reverse.—Within a wreath, " For services in the Mysore, 1791–1792;" surrounded by a Persian inscription,—"A token of the bravery of the troops of the English Government in the war in Mysore, in the year of the Hegira, 1205–1206."

There seem to have been two sizes of this medal in silver.

Worn from a yellow silk cord.

2.—Large size.

3.—Small size.

CEYLON, 1795–1796.

4.—This medal was granted by an Order in Council by the Vice-President, Military Department, Fort William, 15th May, 1807.

Obverse.—"For services on the Island of Ceylon, A.D. 1795–6."

Reverse.—In Persian, " This medal was given by way of acknow-ledgment of services in Ceylon, in the year of the Hegira, 1209–1210."

Worn from a yellow silk cord.

SERINGAPATAM, 1799.

This medal was granted by a general order, Madras, 18th July, 1808, for services in Mysore. The British commanded by General Harris, General Baird, and Colonel Wellesley, who was made Governor of Mysore after the capture of Seringapatam. The siege lasted from the 4th April to 4th May, 1799, on which day the town was taken by storm, and Tippoo, who commanded the enemy, was killed.

Obverse.—The British lion trampling on the Mysorean tiger; above, a pennon, on which is Tippoo's title, " Asadullahal Ghalib, the conquering tiger of God, IV. May, MDCCXCIX."

Reverse.—A storming party advancing to the attack of the town, under the full rays of the sun. Persian inscription : " Seringapatam, God-given, 28th day of the month Likadah, 1218 of the Hegira."

5, 6, 7, 8, 9.—This medal was given in gold, silver, bronze-gilt, bronze, and tin, to the different ranks.

Worn from a buff riband.   Supposed to represent tawny colour of the tiger's skin.

EGYPT, 1801–1802.

10.—This medal was granted by general order, 31st July, 1802.

Major-General Sir David Baird was sent in command of a division numbering 2,800 British troops, 2,000 sepoys, 450 of the Company's picked artillerymen, from India, to assist Sir Ralph Abercromby's force coming from England to act against the French. He was joined on the 17th May, 1801, at Jeddah, on the Red Sea, by an expedition from the Cape of Good Hope, consisting of the 61st regiment, some light horse and artillery, and landed on the 8th June at Cossier.

Obverse.—A Sepoy holding a Union Jack in his right hand; in rear, a camp.   Persian inscription: "This medal has been presented in commemoration of the defeat of the French armies in the kingdom of Egypt, by the great bravery and ability of the victorious army of England.

Reverse.—A ship in full sail with the Union Jack flying; with the pyramids and obelisk in the back ground.   MDCCCI.

Worn like the preceding ones.

A clasp for Egypt was added in 1850 to the war medal, of which Her Majesty sanctioned the grant in 1848.

RODRIGUES, BOURBON, AND ISLE OF FRANCE.
1809, 1810.

11.—General order.   Fort William, 10th September, 1811.   His

Excellency was pleased to signify his approbation of the distinguished merits of the volunteers of the three Presidencies, etc.

In July, 1810, a small Anglo-Indian force under Colonel Keating, and a small squadron of His Majesty's ships under Commodore Rowley, subdued the French in the island of Bourbon.

On the 2nd December, 1810, the French Governor and Commander-in-Chief, De Caen, surrendered the Isle of France to Major-General J. Abercromby, commanding the Army, and Vice-Admiral Bertie, commanding the fleet.

Obverse.—A Sepoy holding the Union Jack, trampling on the French colours and Eagle; by his side a field gun, in the background ships at anchor.

Reverse.—Within a wreath; Persian inscription,—" This medal was conferred in commemoration of the bravery and accustomed fidelity exhibited by the Sepoys of the English Company in the capture of the Mauritius Islands in the year of the Hegira, 1223." Round the wreath, Rodrigues vi. July, mdcccix. Bourbon viii. July, and Isle of France, iii. December, mdcccx.

Worn like the preceding.

Her Majesty's Regiments 69th and 86th Foot have " Bourbon " on their colours.

JAVA, 1811.

12.—This medal was conferred by general order, 11th February, 1812.

The troops, numbering about 12,000, were under the command of Sir S. Auchmuty. Batavia, called by the Dutch " The Queen of the East," was surrendered on the 8th of August, 1811. On the 26th August, " Cornelis," to which fort the Dutch had retired, was taken by assault, under the direction of Colonel Gillespie, Colonel Gibbs, Lieut.-Colonel MacLeod, who was killed, and Major Tule. The final capitulation was signed by General Jansens, commanding the Dutch and Sir S. Auchmuty on the 18th September..

Obverse.—An assaulting party carrying Fort Cornelis; a standard with the British colours over the Dutch; above, " Cornelis."

Reverse.—Persian inscription: " This medal was conferred in commemoration of the bravery and courage exhibited by the Sepoys of the English Company in the capture of the kingdom of Java, in the year of the Hegira, 1228. Round this inscription, " Java conquered xxvi. August MDCCCXI.

Worn like the preceding.

" Java " was one of the clasps given with the War Medal in 1848.

Her Majesty's Regiments 14th, 59th, 69th, 78th and 89th foot, bear it on their colours.

NEPAUL, 1814–1816

13.—Granted by general order.   Fort William, March 20th, 1816.

The first campaign, planned by the Governor-General Lord Moira himself, proved disastrous to the British arms.   The troops were divided into four divisions, the principal under Major-General Marley, consisted of 8,000 men, who signally failed; the 2nd under Major-General Wood, who also failed; the 3rd under Major-General Gillespie, who was killed on the 30th October, 1814; the 4th under Major-General Ochterlony, who gained brilliant successes.

The entire management of the second campaign was left to Major-General Ochterlony, who had 20,000 men, including the 24th, 66th, 87th of His Majesty's Army. On the 27th February, by extraordinary forced marches, he arrived at the town of Mukwaupoor, where he completely routed the enemy.   His brigadiers, Colonel Nicol and Colonel Kelly also, who had been detached, gained signal victories over the enemy, when the Rajah of Nepaul signed the required treaty.

Obverse.—Hills crowned with stockades ; on the left, a field gun.

Reverse.—Persian inscription : " This medal was conferred by the Nawab Governor-General Bahadur, in testimony of the energy, good service, skill, and intrepidity which were displayed during the campaigns in the hills in the years of the Hegira, 1229-1230."

Worn like the preceding.

A clasp for " Nepaul," was granted with the " Army of India" Medal in 1851.

---

lot BURMESE MEDAL, 1824–1826.

14.—Granted by general order.   Fort William, 22nd April, 1826.

Major-General Sir Archibald Campbell commanded the expedition which, in 1824, took the town of Rangoon, and proceeded up the Irrawaddy River ; and, finally, after surmounting all difficulties, concluded peace with the King of Ava, 1826.   The British losses were very great in this campaign, from sickness.   The troops were greatly aided by the men of the fleet, under Sir James Brisbane and Captain Chads.

The troops under General Morrison, that advanced through the Province of Arracan towards the Burmese capital, appear to have received the medal also.

Obverse.—The white elephant of Burmah crouching to the British lion ; the Burmese colours lowered to the Union Jack ; palm-trees in the background. In Persian : " The elephant of Ava submitting to the British lion, 1826."

Reverse.—A storming party advancing to the great Pagoda of Rangoon ; a steamboat on the Irrawaddy ; Sir A. Campbell directing the movement, dismounted under a palm-tree. In Persian : " The standard of the victorious army of England, in Ava."

Riband : Crimson, with blue edge.

A clasp for " Ava " was given with the " Army of India' Medal in 1851.

Her Majesty's Regiments, 1st, 13th, 38th, 41st, 44th, 45th, 47th, 54th, 87th and 89th Foot have " Ava " on their colours.

GHUZNEE, 1839.

> " *General Order,*
>> " *Camp Buddee,*
>>> " *November 23rd, 1842.*

" The Governor-General, being informed that the medals once intended to be given in the name of the late Shah Soojah to the officers and soldiers engaged in the capture of Ghuznee, in 1839, have been manufactured in the Government Mint at Calcutta, and considering it is not just that, in consequence of the death of Shah Soojah, the glorious achievement of the capture by assault of Ghuznee should remain without due commemoration by the conferring of a personal decoration upon those engaged therein, is pleased to direct that the medals prepared shall be given in the name of the Government of India to the officers and men entitled thereto for such service."

The British troops were commanded by Sir John Keane, supported by Brigadier Sale. The enemy by Hyder Khan, one of the sons of Dost Mohamed.

Obverse.—The fortress of Ghuznee, with " Ghuznee " below.

Reverse.—The name and rank and regiment of the recipient. Above, the date, 23rd July; below, a mural crown and the year 1839. The whole surrounded by two laurel branches.

Riband : Crimson and green.

15.—VALENTINE CULLEN ... 16th Dragoons.
> CASE XII., No. 6.

16.—J. ROFE, Corporal ... 2nd Queen's.
> CASE XII., No. 9.

17.—J. ROUND ... ... P.A.L.I.
> CASE XII., No. 11.

18.—JOHN WILSON, Bugler... 13th Light Infantry.
> CASE XII., No. 12.

19.—H. BURROWS, Sergeant 3 Troop, Bombay Horse Artillery.

Her Majesty's Regiments 4th and 16th Light Dragoons, 2nd, 13th and 17th Foot have " Ghuznee " on their colours.

JELLALABAD, 1842.

" General Order,
" Allahabad,
" 30th April, 1842.

" The Governor-General is pleased to direct that a silver medal be made for every officer, non-commissioned officer, and private, European and Native, who belonged to the garrison of Jellalabad on the 7th April, 1842."

General Sale commanded the troops through the gallant defence at Jellalabad from 12th November, 1841, and defeated Akbar Khan in the open field with great loss on the 7th April, 1842, and was joined on the 16th April by General Pollock and his force, when an advance was made towards Cabul.

Obverse.—" Jellalabad " and a mural crown.

Reverse.—" VII. April, 1842."

Riband : Crimson, yellow, and blue, representing the Eastern sky at sunrise.

 20.—T. MANNAN, Sergeant ... ... 13th, or P.A.L.I.
 21.—J. ROUND ... ... 13th Foot.
CASE XII., No. 11.

Her Majesty's 13th Regiment of Foot is the only regiment that has " Jellalabad " on their colours.

" London Gazette,
" 26th August, 1842.

" In consideration of the distinguished gallantry displayed by the 13th Light Infantry during the campaigns in the Burmese Empire and in Afghanistan, Her Majesty has been graciously pleased to approve of that regiment assuming the title of the 13th or Prince Albert's Regiment of Light Infantry, and of its facings being changed from yellow to blue." . . . and to bear on its colours " a mural crown, superscribed Jellalabad."

JELLALABAD, 1842.—2nd Medal.

General order—Commander-in-Chief, 18th March, 1845—notified that new medals had arrived from England, and would be issued to officers and men on their returning those originally presented.

It appears that the Governor-General was dissatisfied with the first issue, and accordingly the second was prepared in England.

Obverse.—Head of Her Majesty, " Victoria Vindex."

Reverse.—A figure of Victory flying over the fortress ; in her right hand a laurel wreath, in her left the Union Jack. " Jellalabad, VII. April, MDCCCXLII."

Riband : The same ; though I was informed by an old soldier, Bugler J. Wilson, 13th Foot, who wore this medal, that he and most of his old comrades wore it with a crimson and blue riband.

22.—ANDREW CULLEN, Private ... 13th Regt. P.A.L.I.
23.—JOHN WILSON, Corporal ... H.M. 13th Foot.

CASE XII., No. 12.

MEDAL FOR CAMPAIGN IN AFGHANISTAN.

General order — Simla, 14th October, 1842 — notified the intention of the Government of India of bestowing a silver medal on the troops engaged in the above campaign.

Obverse.—Head of Her Majesty, " Victoria Vindex."

Reverse.—" Candahar " { Within a wreath surmounted by a crown, 1842.

or

" Ghuznee-Cabul," { Within two wreaths surmounted by a crown, 1842.

or

" Candahar, Ghuznee, Cabul," { Within one wreath surmounted by a crown, 1842.

or

" Cabul." ... ditto.

According to the actions in which the recipient had been engaged.

Riband: The Indian pattern.

General Nott completely defeated the Afghans under the walls of Candahar, and marched on the 15th August, 1842, to Ghuznee, which he found evacuated, and proceeded to Cabul.

On the 20th August, Generals Pollock and Sale advanced from Jellalabad to Cabul; about the middle of September, the city was in the hands of the British—and the English prisoners rescued.

Her Majesty's 40th and 41st Regiments have "Candahar" on their colours, 3rd Dragoons, 9th, 13th, 31st, 40th and 41st, "Cabool, 1842." 40th and 41st, "Ghuznee;" in addition to those regiments who received the distinction at the first capture, 1839.

CANDAHAR.

24.—MICHAEL SULLIVAN, Private ... H.M.'s 40th Regiment.

GHUZNEE, CABUL.

25.—JAMES CARTER, Gunner ... 1st Troop, Horse Brigade, Bombay Artillery.

CANDAHAR, GHUZNEE, CABUL.

26.—THOS. HOPKINS ... ... 41st Regiment.

CABUL.

27.—WILLIAM HULL, Sergeant ... Bandmaster, 3rd K.O.L.D.
28.—W. ROFE ... ... 2nd Queen's Royals.
CASE XII., No. 9.

29.—J. Round ... ... 18th P.A.L.I.
Case XII., No. 11.

30.—John Wilson, Corporal ... 13th Foot.
Case XII., No. 12.

31.—Alexander Smith, No. 928 ... H.M. 31st Regiment.
Case XII., No. 14.

CABVL, 1842.

32.—George Albury ... ... 3rd Dragoons.
Case XII., No. 2.

KELAT-I-GHILZIE, 1842.

" General Order,
" Simla,
" October 4th, 1842.

" To every officer, non-commissioned officer, and private present within Kelat-I-Ghilzee, and forming part of the garrison thereof during the late investment and blockade of that fort, will be presented a Silver Medal, etc."

The garrison consisted of 600 of Sheik's Army, three company's Bengal 43rd Native Infantry, under Captain Webster and Lieutenant Trotter, forty European Artillery, and sixty Sappers and Miners—about 950 in all—under Captain John Halkett Craigie.

Obverse.—A shield, on which is " Kelat-I-Ghilzee," surrounded by laurel branches, surmounted by a mural crown.

Reverse.—A trophy of arms. " Invicta, MDCCCLXIII."

Riband: Indian pattern.

33.—Matthew Bradley, Gunner ... 4th Company, 2nd Battery, Artillery.

MEDAL FOR THE SCINDE CAMPAIGN, 1843.

General order, 22nd September, 1843, announced the intention of the Court of Directors, dated 2nd August, 1843, of bestowing a Medal for this campaign.

Sir Charles Napier crossed the Indus and completely routed the enemy in the great battle at Meeanee, on the 17th February, 1843. The Beloochees fought most gallantly, leaving no less than 6,000 dead on the field. He again defeated the enemy at Dubba, 24th March, 1843, near Hyderabad. On the 18th March, 1843, the Governor-General appointed, in a proclamation, Major-General Sir Charles Napier, K.C.B., governor of the province he had conquered.

Obverse.—Head of Her Majesty. " Victoria Regina."

Reverse.—" Meeanee, 1843 ; "

or

" Meeanee, Hyderabad, 1843 ; "

or

" Hyderabad, 1843."

According to the actions in which the recipient had been engaged.

Surrounded by laurel branches, surmounted by a crown.

Riband : Indian pattern.

Her Majesty's Regiment, 22nd Foot, is the only one that has " Scinde," " Meeanee," and " Hyderabad " on their colours.

MEEANEE, 1843.

34.—J. L. Pennefather, Lieut.-Col.   22nd Regiment.

MEEANEE, HYDERABAD.

35.—Jas. Townsend...   ...   ...   22nd Regiment.

HYDERABAD.

36.—John Bogle, Corporal ...   ...   22nd Regiment (2).

MEDALS FOR THE CAMPAIGN IN GWALIOR, 1843.

" General Order, 4th January, 1844.

" The Government of India will, as a mark of its grateful sense of their distinguished merits, present to every general and other officers, and to every soldier engaged in the battles of Maharajpoor and Punniar, an Indian star of bronze, made out of the guns taken at those battles, and all officers and soldiers in the service of the Government of India will be permitted to wear the star with their uniforms."

Sir Hugh Gough advanced early in December from Agra ; at the same time another division, under Major-General Grey, advanced from Bundelkund.   They both encountered and defeated the enemy on the same day, 29th December, 1843 ; the former at Maharajpoor, the latter at Punniar.

Obverse.—A bronze star of six points, with a smaller one of silver on it, with " Maharajpoor, 29th Dec., 1843," or " Punniar, 29th Dec., 1843."

Reverse.—The recipient's name, rank, and regiment.

Riband : Indian pattern.

These medals were made from the guns captured in the two battles.

Her Majesty's Regiments, 16th Dragoons, and 39th and 40th Foot have " Maharajpoor " on their colours ; and the 9th Dragoons, and 3rd and 50th Foot, " Punniar."

MAHARAJPOOR.

37.—V. Cullen, Sergeant ...   ...   H.M. 16th Lancers.
Case XII., No. 6.

38.—E. John Hopewell, Private ...   H.M. 39th Regiment.

39.—Thomas Johnson, Sergeant   ...   H.M. 16th Lancers.
Case XII., No. 7.

PUNNIAR.

40.—John Williams, Corporal   ...   50th Queen's Own Regt.

MEDAL FOR SUTLEJ, 1845, 1846.

General order, 17th April, 1864. This order granted the issue of Medals to the troops engaged at Moodkee in addition to those granted to the men engaged at Ferozeshuhur, Aliwal, and Sobraon.

Sir Hugh Gough defeated the Sikhs at Moodkee on the 18th December, 1845. On the 21st December, he advanced and met the force coming from Ferozpoor under Sir John Littler, numbering about 5,500 men, with 21 guns. The combined forces under Brigadiers Major-General Gibbert, Sir John Littler, Brigadier Wallace, and Sir Harry Smith, attacked the entrenched camp at Ferozeshuhur on the afternoon of that day, but darkness coming on, they were unable to complete their victory till the next day, when the camp was taken. The enemy fled across the Sutlej. The army was unable to follow the enemy, owing to its weakness in Cavalry. But Sir Harry Smith, who had been detached, again defeated the enemy at Aliwal, on the 28th January, 1846. The Sikhs having again crossed the Sutlej, were defeated by Sir Hugh Gough, with terrible slaughter, on the 10th February, 1846, at Sobraon.

Obverse.—Head of Her Majesty. " Victoria Regina."

Reverse.—A standing figure of Victory holding a laurel wreath in her out stretched hand ; a trophy of Indian Arms. " Army of the Sutlej."

Moodkee,   Ferozeshuhur,   Aliwal,   or   Sobraon,
1845.   1845.   1846.   1846,

whichever being the first action in which the recipient was engaged ; for each subsequent action he received a clasp.

Riband : Blue, with crimson border.

The following Regiments bear the several distinctions on their colours :—

Moodkee—3rd Light Dragoons ; 9th, 31st, 50th and 80th Foot.
Ferozeshuhur—3rd Light Dragoons ; 9th, 29th, 31st, 50th, 62nd
and 80th Foot.
Aliwal—16th Lancers ; 31st, 50th and 53rd Foot.
Sobraon—3rd Light Dragoons ; 9th and 16th Lancers ; 9th, 10th,
29th, 31st, 50th, 53rd, 62nd and 80th Foot.

MOODKEE, 1845.

41.—James Thrift ...　　...　　... 80th Regiment.

FEROZESHUHUR, 1845.

42.—Michael Byrne　　...　　... 62nd Regiment.

ALIWAL, 1846.

43.—Wm. Turner ...　　...　　... 16th Lancers.

SOBRAON, 1846.

44.—Daniel Maloney　　...　　... 10th Regiment.

Medal and One Clasp.

MOODKEE, 1845; FEROZESHUHUR.

45.—Arthur Hamill ... ... 50th Regiment.

MOODKEE, 1845; SOBRAON.

46.—T. Blak, Rough Rider... ... 1st Brigade H.A.

FEROZESHUHUR, 1845; ALIWAL.

47.—T. Cronin, Sergeant ... ... 4th Battalion Artillery.

FEROZESHUHUR, 1845; SOBRAON.

48.—George Phillips ... ... 29th Regiment.

ALIWAL, 1846; SOBRAON.

49.—Thomas Johnston, Sergeant ... 16th Lancers.

Case XII., No. 7.

50.—Wm. Fardell ... ... 62nd Regiment.

Medal and Two Clasps.

MOODKEE, 1845; FEROZESHUHUR, SOBRAON.

51.—John Lindsay .. .. K.O. 3rd Lt. Dragoons.

MOODKEE, 1845; ALIWAL, SOBRAON.

52.—John Shelley ... ... 31st Regiment.

FEROZESHUHUR, 1845; ALIWAL, SOBRAON.

53.—Henry Allward ... ... 62nd Regiment.

Medal and Three Clasps.

MOODKEE, 1845; FEROZESHUHUR, ALIWAL, SOBRAON.

54.—Alexander Smith, Sergeant ... 31st Regiment.
Case XII., No. 14.

55.—James McGool... .. . 50th Regiment.

PUNJAB MEDAL, 1848—1849.

" General Order,
> *" Camp, Ferozepore,*
>> *" 2nd April, 1849.*

"In anticipation of the wishes of the Honourable Court of Directors, the Governor-General will grant to every officer and soldier who was employed with the Punjab in this campaign, to the date of the occupation of Peshawar, a Medal, etc.

Lord Gough, the Commander-in-Chief, personally took command in this campaign. After a prolonged and arduous siege, from the 7th September, 1848, to 2nd January, 1849, Moolraj, the Sikh Chief surrendered Mooltan to Major-General Whish. The British loss was very severe, numbering 1,200 killed and wounded during the siege. Lord Gough defeated the Sikhs with great slaughter, on the 13th January, 1849, at Chilianwala. The British loss was very great, especially amongst the 24th Foot. After their defeat, the enemy quitted their entrenchments, and took up a position between Goojerat and Chenab. Having been joined by Major-General Whish, Lord Gough again defeated the enemy on the 21st February, 1849. The result of these victories was the annexation of the Punjaub.

Obverse.—Head of Her Majesty. "Victoria Regina."

Reverse.—Sikhs laying down their arms to Lord Gough, who is mounted; the British Army in line; in background, palm-trees. MDCCCXLIX.

Clasps for Mooltan, Chilianwala, Goojerat.

Riband: Dark blue, yellow edges.

The following Regiments have the several distinctions on their colours :—

Mooltan—10th, 32nd and 60th Foot.

Chilianwala—3rd, 9th and 14th Light Dragoons; 24th, 29th and 61st Foot.

Goojerat—3rd, 9th and 14th Light Dragoons; 10th, 24th, 29th, 32nd, 53rd, 60th and 61st Foot.

MOOLTAN.

56.—W. Cowie ... ... 1st Batt. 60th R. Rifles.

CHILIANWALA.

57.—J. Collins ... ... 3rd Light Dragoons.

GOOJERAT.

58.—George Frary ... ... 53rd Foot.

𝕸𝖊𝖉𝖆𝖑 𝖆𝖓𝖉 𝕿𝖜𝖔 𝕮𝖑𝖆𝖘𝖕𝖘.

MOOLTAN, GOOJERAT.

59.—A. Smith, Sergeant ... ... 32nd Foot.
Case XII., No. 14.
60.—James O'Donnell ... ... 1st European Fusiliers.

CHILIANWALA, GOOJERAT.

61.—J. Bancroft ... ... 24th Foot.

"ARMY OF INDIA MEDAL, 1799–1826."

General order, 21st March, 1851, notified that Her Majesty had given her royal consent to the Directors of the East India Company for granting a medal at their expense to the surviving officers and soldiers who had been engaged in India from 1799 to 1826.

Obverse.—Head of Her Majesty. "Victoria Regina."

Reverse.—A figure of Victory, seated, holding in her right hand a laurel branch, in her left a wreath, at her feet a trophy of Indian arms; in the background, palm-trees. "1799–1826."

Riband: Light blue.

The following clasps were issued with this medal:—

Allighur ...	... ...	Siege of, 4th September, 1803
Battle of Delhi ...	... ...	11th ,, ,,
Assaye ...	... ...	23rd ,, ,,
Asseerghur ...	... ...	Siege of, 21st October, ,,
Laswarree ...	... ...	1st November, ,,
Argaum ...	... ...	29th ,, ,,
Gawilghur... ...	Siege and storming of, 15th December, ,,	
Defence of Delhi ...	... ...	October, 1804
Battle of Deig ..	... ...	13th November, ,,
Capture of Deig ...	... ...	23rd December, ,,
Nepaul ...	... ...	... 1816
Kirkee ...	... ...	{ Capture } November, 1817
Kirkee and Poona...	... ...	{ Battle }
Seetabuldee ...	... ...	{ November and }
Nagpore ...	... ...	{ December ... } ,,
Maheidpore ...	... ...	21st December, ,,
Corygaum...	... ...	1st January, 1818
Ava ...	... ...	... 1824–1826
Bhurtpore...	... ...	... Siege of, January, ,,

But few of these actions are borne on the colours of Her
Majesty's regiments, though, with one or two exceptions, some
British regiments took part in each engagement.

ALLIGHUR.

62.—The garrison of Allighur, the head quarters and military
depôt of M. Perron, the French agent, surrendered to General Lake
on the 4th September, 1803. The 27th Dragoons and 76th Foot
were present, but do not bear the distinction on their colours.

BATTLE OF DELHI.

63.—General Lake defeated the Mahrattas under Louis
Bourquien, who had succeeded M. Perron, on the 11th September,
1803. 27th Light Dragoons and 76th Foot present; not on colours.

ASSAYE.

64.—This was General Wellesley's first victory. He defeated the
Mahratta Army under their Chief Scindiah, on the 23rd September,
1803. The British loss was most severe, 22 officers and 386 men
killed, and 57 officers and 1,526 men wounded.

19th Light Dragoons, 74th and 78th Foot. The two latter have
Assaye on their colours.

M. Jordan ... ... 19th Light Dragoons.

ASSEERGHUR.

65.—This fortress surrendered to Colonel Stevenson on the 21st October, 1803. His Majesty's 94th Regiment was present, though they do not bear the distinction.

H. McLeod ...　　...　　...　　... 94th Foot.

LASWARREE.

66.—General Lake defeated the Mahrattas on the 1st November, 1803. The brunt of the battle was born by His Majesty's 76th Regiment. "This handful of heroes" as they were called by Lake, repulsed the Mahratta horse. The 27th, 29th and 8th Light Dragoons were present. The latter only bear the distinction of "Laswarree" on their appointments. It is supposed that 7,000 of the enemy were left dead on the field. The seventeen disciplined battalions known as the "Deckan Invincibles" being utterly destroyed. General Lake was created a peer after this action.

ARGAUM.

67.—General Wellesley, having effected a junction with Colonel Stevenson, here completely routed the Mahrattas under Scindiah, and the Rajah of Berar, on the 29th November, 1803. 19th Light Dragoons, 74th, 78th and 94th, His Majesty's Regiments, were present.

GAWILGHUR.

68.—General Wellesley lost no time in laying siege to Gawilghur, one of the strongest fortresses in India, on the 15th December, 1803; it was taken by storm. The light companies of His Majesty's 94th Regiment, under Captain Campbell, led the assault, behaving most gallantly. The 74th, 78th, and 94th, His Majesty's Regiments, were present.

DEFENCE OF DELHI.

69.—This gallant defence was made by Colonel Ochterlony and Lieutenant-Colonel Burn. They had none of His Majesty's troops under their command. Holkar raised the siege on the approach of General Lake's army, on the 15th October, 1804.

BATTLE OF DEIG.

70.—General Frazer marched in pursuit of Holkar, and defeated him on the 13th November, 1804. The British General was mortally wounded. His Majesty's 76th Regiment, the only British regiment present, behaved most gallantly.

CAPTURE OF DEIG.

71.—Lord Lake, having been joined by his reserve under Colonel Don, and a battering train from Agra, laid siege to Deig. On the morning of the 24th December, 1804, the British were in possession of the outer works, and on the morning of Christmas Day the Mahrattas evacuated the fortress. His Majesty's Regiments engaged were the 8th Light Dragoons, 22nd Regiment (flank companies) and the 76th Foot.

NEPAUL.

72.—*Vide* the Nepaul Medal, granted to the East India Company's troops. The divisions under Generals Marley and Wood were excluded from this Medal, having " failed to penetrate into the hills." General order, 13th February, 1852. His Majesty's 8th Light Dragoons, 14th, 17th, 24th, 66th and 87th Foot, were engaged in this campaign.

W. Baron ... ..: ... ... 17th Foot.

KIRKEE.

73.—On Mr. Elphinstone discovering that the Peishwa's agents were attempting to corrupt the Company's Sepoys that were at Poonah, he removed them, consisting of one brigade, to the village of Kirkee. It was determined not to await the arrival of reinforcements ; but the brigade under Colonel Burr, about 2,800 men, at once attacked the Mahrattas, and routed them. They fled to the town of Poonah, November, 1817. None of His Majesty's Regiments were present.

POONAH.

74.—Colonel Burr waited for the arrival of Brigadier-General Lionel Smith and his troops, and after repeated skirmishes with the Mahrattas, they prepared to lay siege to Poonah, which they found almost evacuated, November, 1817. His Majesty's 65th Regiment was present.

    I. BRINKWORTH   ...   .   65th Foot.

KIRKEE AND POONAH.

75.—The Medal with this clasp was given to men who had been present at both actions.

    GEORGE McBRIDE, Sergeant ...   ...   European Regiment.

SEETABULDEE.

76.—Apa Sahib, at Nagpore, not knowing of the defeat of the Peishwa at Poonah, openly declared for him. Lieutenant-Colonel H. S. Scott, commanding at Nagpore, was attacked by the Arabs when posting sentries on the hill of Seetabuldee, which overlooked the Residency, on the 26th November, 1817. The action lasted till 12 o'clock on the next day. None of His Majesty's Regiments were present.

NAGPORE.

77.—Reinforcements arrived, under the command of Brigadier-General Doveton. The result of the treachery evinced by the Rajah resulted in the Battle of Nagpore, on the 16th December, 1817. The Arabs refused to evacuate the city. The assault, on the 23rd September, proved unsuccessful, but the enemy surrendered on the 30th December. His Majesty's 1st Foot was present, and have " Nagpore " on their colours.

P. KELLY ... ... 1st Foot.

SEETABULDEE, NAGPORE.

78.—A clasp inscribed thus was given to those who had been present in both actions.

MAHEIDPOOR.

79.—The Patan chiefs, after fearful atrocities committed in Holkar's camp, became clamorous for battle, and on the 21st December, 1817, they met their reward in the decisive battle at Maheidpoor. The British were commanded by Sir Thomas Hislop and Sir John Malcolm. A part of His Majesty's 22nd Light Dragoons, and the flank companies 1st Foot, were present; the latter have " Maheidpore " on their colours.

JOHN O'BRYAN, Gunner... ... ... Artillery.

CORYGAUM.

80.—This gallant defence was made by Captain Staunton, who was proceeding to Soroor for reinforcements with about 800 men and 2 guns; on being surrounded by the whole army of Mahrattas, consisting of about 25,000 men, he made a dash at the village of

Corygaum, and successfully held it against the repeated attacks of the enemy, who withdrew on the approach of General Smith's division.

A V A.

81.—*Vide* the account of the Medal given to the East India Company's troops for the war in Burmah, 1824–1826.

J. PIGEON ... ... 87th Foot.

B H U R T P O O R.

On the 10th December, 1825, in consequence of the hostility shown by the people of Bhurtpoor, Lord Combermere, with an army of 20,000 men, sat down before the place; on the 18th January, 1826, the city was surrendered. His Majesty's 11th Light Dragoons, 16th Lancers, 14th and 59th Regiments took part in the siege, and all bear the distinction on their appointments.

82.—J. BRADSHAW, Troop Sergt.-Major 11th Light Dragoons.
CASE XII., No. 4.

83.—T. LOWE ... ... 16th Lancers.

84.—J. LORD, Sergeant ... ... 59th Foot.
CASE XII., No. 20

Two Clasps.

ALLIGHUR, LASWARREE.

85.—N. RICHMOND ... ... 29th Light Dragoons.

ASSAYE, ARGAUM.

86.—D. McLEOD ... ... 78th Foot.
CASE XII., No. 22.

LASWARREE, CAPTURE OF DEIG.

87.—J. TRAVERS ... ... 8th Light Dragoons.

NAGPORE, AVA.

88.—T. REDDEN, Sergeant ... ... 1st Foot.

NEPAUL, BHURTPOOR.

89.—T. HOWE ... ... 14th Foot.

MAHEIDPOOR, AVA.

90.—H. LEE ... ... 1st Foot.

Three Clasps.

ASSAYE, ARGAUM, GAWILGHUR.

91.—John Cameron ...   . 78th Foot.

LASWARREE, CAPTURE OF DEIG, NEPAUL.

92.—P. Reilly, Corporal ...   ... 8th Light Dragoons.

Four Clasps.

ALLIGHUR, BATTLE OF DELHI, LASWARREE, BATTLE OF DEIG.

93.—P. King   ...   . 76th Foot.

BURMAH, 1852–1853.

———

General order, 23rd January, 1854, notified that—

" The Queen has been pleased to sanction the issue of a Medal for the purpose of commemorating the services rendered during the operations against the Burmese."

After ineffectual attempts at negociation, hostilities were commenced against the King of Ava, who refused redress for injuries inflicted on British subjects at Rangoon. Major-General Godwin embarked with his forces on the 28th March, 1853, and speedily destroyed the stockades on the Irrawaddy. Rangoon was captured on the 14th April. The important city of Prome was taken in October. On the 21st November, Pegu was attacked by Brigadier McNeill. After three days' fighting the stronghold of Myat-toon, a robber chief, was reduced by the troops under Sir John Cheape, K.C.B., on the 19th March, 1853. On the 30th June, 1853, the termination of the war was officially announced. The 18th, 51st and 80th Regiments have " Pegu " on their colours.

Obverse.—Head of Her Majesty. " Victoria Regina."

Reverse.—A figure of Victory crowning a classic warrior, seated, with a laurel wreath ; below, lotus leaves.

This medal was issued with a clasp for Pegu.

Riband : Crimson, with two dark blue stripes.

94.—Patrick Green, Private　　 ...　80th Regiment.

———

PERSIA, 1856–1857.

General order, 12th April, 1858, notified the grant of this Medal.

War was declared on the 1st of November, 1856, in consequence of Persia having taken possession of Herat. The troops under Major-General Stalker, C.B., were landed at Bushire on the 7th November, and two days afterwards the fort of Reshire was taken. A larger force subsequently arrived from Bombay, under Lieutenant-General Sir James Outram. After repeated successes, a treaty of peace was signed at Bagdad on the 2nd May, 1857.

The medal is the same as that issued for the war in Burmah, with a clasp "Persia." Her Majesty's 14th Light Dragoons, 64th and 78th Foot, have "Persia" on their colours. The 64th also have "Bushire, Koosh-ab, Reshiher"; and the 78th, "Koosh-ab."

95.—C. Woollett ...	...	...	14th King's Lt. Dragns.
96.— McKay ...	...	...	78th Highlanders.

Case XII., No. 23.

This medal is now known as the "India General Service Medal," and has been granted by Her Majesty with the following clasps :—

North-West Frontier.	Looshai.
Umbeyla.	Jowaki.
Bhootan.	Perak.

NORTH-WEST FRONTIER.

General order, 1st July, 1869. Her Majesty granted the issue of the Medal and clasp for the troops who had been engaged on the North-West frontier from the year 1849 to 1868.

97.—B. Stubbs, Corporal, 1,494... 1st Batt. H.M. 19th Reg.

UMBEYLA.

General order, 1st July, 1869. This clasp was granted by Her Majesty to the troops who had been present at Umbeyla.

    98.—W. H. Owen, 3,490 ...   ... 93rd Highlanders.

BHOOTAN.

General order, 29th of April, 1870. Her Majesty granted the issue of the Medal and clasp to the troops engaged in the years 1864-65-66.

    99.—C. Arver, 218 ...   . H.M. 80th Regiment.

LOOSHAI.

100.—General order, December, 1872. Her Majesty granted the issue of this clasp.

JOWAKI, 1877–78.

101.—General order, 1st March, 1879. Her Majesty granted this Medal to the troops engaged against the Afridis, 1877–78.

PERAK, 1875–1876.

102.—General order, 1st September, 1879. Her Majesty granted the issue of this clasp.

MERITORIOUS SERVICE.

———

Instituted by general order, 20th May, 1848.

Obverse.—Head of Her Majesty. " Victoria Regina."

Reverse.—Company's Arms. "Auspicio Regis et Senatus Angliæ "
On a scroll, " For Meritorious Service."

Riband : Crimson.

    103.—H. WATKINS, Sergeant   ...   BombayHorse Artillery.

———

LONG SERVICE AND GOOD CONDUCT.

———

Instituted by general order, 20th May, 1848.

Obverse.—A shield with East India Company's Arms, and a
trophy of arms.

Reverse.—In the centre, recipient's name, etc. " For long service
and good conduct."

Riband : Crimson.

    104.—H. A. GILBERT, Sergt.-Major... Sappers and Miners.
                                         CASE XII., No. 36.

    105.—V. B. CULLEN, Sergt.-Major... 8th Bengal Cavalry.
                                         CASE XII., No. 6.

    106.—JAS. ENGLISH, Qrmaster. Sergt. European Invalid Batt.
       17th May, 1855.

———

LONG SERVICE AND GOOD CONDUCT (2.)

———

    This Medal was issued by mistake, probably intended for a naval
medal, but the East India Company never gave a good conduct
medal to their navy.

N

Obverse.—Head of Her Majesty. " Victoria Regina."

Reverse.—" For long service and good conduct," surrounded by two oak branches ; above, a crown ; below, an anchor.

Riband : Crimson.

107.—Matthew Quick, Staff-Sergt.... Gwalior Contingent.
30th April, 1858.

CASE IV.

INDIA. MUTINY.

SOUTH AFRICA.	1ST CHINA.
2ND CHINA.	NEW ZEALAND.
ABYSSINIA.	ASHANTEE.

SOUTH AFRICA.

General order, 22nd November, 1854.   Her Majesty granted this Medal to commemorate the successes of Her Majesty's forces in the war in which they had been engaged against the Kaffirs in the years 1834–5, and 1846–7, and between the 24th December, 1850, and the 6th February, 1853.

Obverse.—Head of Her Majesty.   " Victoria Regina."

Reverse.—A lion crouching under a shrub; " South Africa " above; below, " 1853."

Riband : Orange, with two broad and two narrow blue stripes.

Major-General Sir Benjamin D'Urban, K.C.B., was the British Commander-in-Chief in the campaign, 1834–35, with Colonel Smith, C.B., second in command.   He defeated the Kaffir chief, Huitza.   Her Majesty's 27th, 72nd and 75th regiments were engaged.

In the second campaign, 1846–47, two small divisions invaded Kaffirland, commanded by Lieutenant-Colonel Robert Richardson, and Colonel H. Somerset.   Sir Peregrine Maitland arrived from Port Victoria, in April, 1846.   Active operations were brought to a close in October, when the chief, Sandilli, surrendered.   Lieutenant-General Sir Harry Smith, Bart., G.C.B., assumed the command and governorship of the colony on the 17th December, 1847.   Her Majesty's 7th Dragoon Guards, 6th, 27th, 45th, 73rd, 90th and 91st regiments were engaged.

In 1850–53, the Kaffirs again showed signs of hostility.   Sir Harry Smith with difficulty escaped from Fort Cox, and established his head-quarters at King William's Town.   Various successful attacks were made on the enemy during this war by Major-General Somerset, Colonel Mackinnon, Lieutenant-Colonel John Michel, and Lieutenant-Colonels Eyre and Percival.   Sir George Cathcart relieved Sir H. Smith in April, 1852 ; Major-General Yorke commanding the second division.   On the 12th March, 1853, peace was concluded, the chiefs Sandilli and Macomo, and the Gaika people, having surrendered.

Her Majesty's 12th Lancers, 2nd, 6th, 12th, 43rd, 45th, Royal Marines, 2nd battalion 60th, 73rd, 74th, 91st and 1st battalion Rifle Brigade, were engaged.

1.—1st campaign	...	J. LAING, 72nd Regiment.
2.—2nd	,,	...	WM. LONG, 90th Regiment.
3.—3rd	,,	...	W. THOMPSON, 91st Regiment.

INDIA.   MUTINY.

This Medal was granted by general order of the Indian Government, No. 383, dated 18th August, 1858, to the troops engaged in the suppression of the mutiny, 1857–58.

Obverse.—Head of Her Majesty. " Victoria Regina."

Reverse.—A helmeted figure of Victory, standing beside the lion ; in her outstretched right hand a laurel wreath ; on her left arm a shield, with the crosses of the Union ; above, "India ; " below, " 1857–1858."

Riband : Scarlet and white stripes.

The following clasps were given with this medal :—

Delhi.	Lucknow.
Defence of Lucknow.	Central India.
Relief of Lucknow.	

DELHI.

This clasp was granted to the troops employed in the operations against and assault of Delhi, 30th May to 14th September, 1857. General the Honourable G. Anson, Commander-in-Chief, died of cholera. The command devolved upon Brigadier-General Reed, who was subsequently unable to proceed through sickness, when Major-General Sir Henry Barnard succeeded to the command.

Her Majesty's 8th, 52nd, 60th, 61st, 75th, 80th, 83rd, 101st and 104th Regiments have "Delhi" on their colours. Also 6th Dragoon Guards and 9th Lancers.

4.—E. Bastable ... ... 1st Batt. 60th Rifles.

DEFENCE OF LUCKNOW.

This clasp was given to the original garrison under Major-General Inglis, and to those who succoured them and continued the defence under Major-Generals Sir Henry Havelock and Sir James Outram, until relieved by Lord Clyde, 29th June to 25th September, 1857.

5.—J. G. W. Harrison ... ... 32nd Light Infantry.

RELIEF OF LUCKNOW.

Was given to the troops engaged against the place, under Lord Clyde, November, 1857.

6.—M. F. Hungerford ... ... 64th Regiment.
7.—J. Munro, Sergeant ... ... 93rd Highlanders.

Case XII., No. 28.

LUCKNOW.

This clasp was given to the troops engaged in the final operations and capture, under Lord Clyde, 2nd to 21st March, 1858.

Her Majesty's 7th Hussars, 9th Lancers, 5th, 8th, 10th, 23rd, 32nd, 34th, 38th, 42nd, 53rd, 64th, 75th, 78th, 79th, 82nd, 84th, 90th, 93rd, 97th, 101st, 102nd, and Rifle Brigade have "Lucknow" on their colours.

8.—R. Lewis, Corporal	...	...	Royal Engineers.

CASE XII., No. 35.

9.—William Hepburn	...	...	2nd Batt. Rifle Brigade.

CASE XII., No. 30.

10.—Andrew Lee	...	...	97th Regiment.

CENTRAL INDIA.

This clasp was given to the troops under Major-General Sir Hugh Rose, G.C.B., engaged in the operations against Shansi, Calpee, and Gwalior; also to the troops under Major-Generals Roberts and Whitlock, January to June, 1858.

Her Majesty's 8th Hussars, 12th Lancers, 14th Hussars, 71st, 72nd, 80th, 83rd, 86th, 88th, 95th, and 108th Regiments have "Central India" on their colours.

11.—Michael O'Brien	... ...	83rd Regiment.

The medal was also given without any clasp.

12.—H. Holmes	...	...	2nd Dragoon Guards.
13.—J. McKay	...	...	78th Highlanders.

CASE XII., No. 23.

14.—H. A. Gilbert, Sergeant-Major	Sappers and Miners.

CASE XII., No. 36.

Two Clasps.

DELHI, RELIEF OF LUCKNOW.

15.—Joseph Batman, Corporal   ...   1st Batt. 8th Regiment.

DELHI, LUCKNOW.

16.—William Brown, Corporal   ...   1st Europ. Beng. Fusl.

DEFENCE OF LUCKNOW, LUCKNOW.

17.—G. Edwards, 2789   ...   ...   H.M's. 81st Regiment.

RELIEF OF LUCKNOW, LUCKNOW.

18.—Henry Cruwys, Sergeant   ...   53rd Regiment.

Three Clasps.

DELHI, RELIEF OF LUCKNOW, LUCKNOW.

19.—John D. Daly   ...   ...   9th Lancers.

CHINA, 1840—1842 (1st).

This Medal was granted by the India Government.

Brigadier-General Burrell, of the 18th Royal Irish, was sent from Madras to the Island of Chusan. On 5th July, 1840, his troops effected a landing. After various successes by the troops under Major-General Sir Hugh Gough, Lieutenant-Colonel Craigie, Major-Generals Lord Saltoun and Schoedde, K.C.B., during the two following years, the City of Nankin was surrounded by the troops under Sir H. Gough, on the 9th August, 1842. Peace was then concluded, the Chinese paying an indemnity and ceding territory.

Obverse.—Head of Her Majesty. " Victoria Regina."

Reverse.—A trophy of Naval and Military weapons, under a palm-tree, with a shield bearing the Royal Arms ; above, " Armis exposcere pacem ; " below, " China, 1842."

Riband : Crimson, with yellow border.

Her Majesty's 18th, 26th, 49th, 55th and 98th Regiments have " China" (with the dragon) on their colours.

  20.—Martin Murray ... ... 18th Regiment of Foot.

CHINA (2nd).

General Order, No. 778, 6th March, 1861. Her Majesty granted the issue of this Medal to the troops engaged in the Chinese War, 1857–60.

Obverse.—The same as the 1st China.

Reverse.—The same, with date 1842 left out.

Riband : The same.

The following clasps were issued with this Medal :—

Canton, 1857.	Taku Forts, 1860.
Taku Forts, 1858.	Pekin, 1860.

CANTON, 1857.

The British troops were landed in two Brigades at the south-east of the town on the 28th December, 1857, assisted by the French Naval Brigade of 900 men. Colonel Holloway, Royal Marines, commanded the 1st Brigade ; Colonel Hope Graham, 59th Foot, commanded the 2nd Brigade ; Colonel Dunlop, the Artillery. On the 29th the city was taken, and on the 5th January, 1858, the Tartar General was captured.

Her Majesty's 59th Regiment is the only one that has " Canton " on its colours.

21.—HY. BRODERICK ... ... 59th Regiment.

TAKU FORTS, 1858.

On the 20th May, 1858, these forts, at the mouth of the Peiho, were captured ; on the 20th June, a treaty was signed.

22.—JOHN JOICE ... ... 1st Battalion, 3rd Foot.

TAKU FORTS, 1860.

The Chinese Government having refused to ratify the treaty, Major-General Sir James Hope Grant, K.C.B., was appointed to command the troops. The French troops were under the command of General Montauban. On the 21st August, 1860, the allied armies captured these forts.

Her Majesty's 1st King's Dragoon Guards, 1st, 2nd, 3rd, 31st, 44th, 60th and 67th Regiments have " Taku forts " on their colours.

23.—ROBERT CHIRM ... ... 44th Regiment.

24.—H. GIBBERT, Sergeant-Major ... Sappers and Miners.

CASE XII., No. 36.

PEKIN, 1860.

———

Two engagements took place on the 18th and 21st September, on the Chow Ho, when the Chinese entrenched camp was taken, and they were driven back on Pekin, which city was peacefully taken possession of by the allied army, on the 13th October, 1860.

Her Majesty's King's Dragoon Guards, 1st, 2nd, 60th, 67th and 99th Regiments have "Pekin" on their colours.

25.—Patrick Walsh ... ... 99th Regiment.

Two Clasps.

TAKU FORTS, 1860 ; PEKIN, 1880.

26.—Daniel Shea ... ... 67th Regiment.
27.—Richard Lewis, Sergeant ... Royal Engineers.

Three Clasps.

CANTON, 1857 ; TAKU FORTS, 1860 ; PEKIN, 1860.

28.—George Lackey, Gunner ... No. 6 B., 12th Brigade, Royal Artillery.

NEW ZEALAND.

General Order 17, of 1st March, 1869. Her Majesty granted the issue of this Medal to the troops who had been engaged in the wars in New Zealand during the years 1845–47, and 1860–66.

Her Majesty's 12th, 14th, 18th, 40th, 43rd, 50th, 57th, 58th, 65th, 68th, 70th, 96th and 99th Regiments have " New Zealand " on their colours.

Obverse.—Head of Her Majesty diademed, with veil covering the back of head and neck. " Victoria, D.G. Britt., Reg., F.D."

Reverse.—Two laurel branches, in which is inscribed the years the recipient was engaged. " New Zealand. Virtutis honor."

Riband : Blue, with scarlet stripe in the centre.

MEDAL WITHOUT DATE.

29.—Joseph Overall, 3089   ...   43rd Regiment.

WITH DATE.

1861.

30.—J. Kelly, 728 ...   .   40th Regiment.

1866.

31.—John McCorry, 1299 ...   ...   1st Batt., 12th Regt.

1860–1864.

32.—G. H. Boggs, 186   ...   ...   40th Regiment.

1860–1865.

33.—John Sands, 220   ...   ...   65th Regiment.

1860–1866.

34.—Owen Williams, 868 ...   ...   1st Batt., 12th Regt.

1861–1866.

35.—Joseph Fisher, 74 ... ...   57th Regiment.

1863–1864.

36.—Wm. Petrie, Sapper, 4290   ...   Royal Engineers.

1863–1865.

37.—Alfred Polton, Dr., 1814 ... 2nd Batt., 18th Royal
Irish Regiment.

1863–1866.

38.—Wm. Adams, 248 ... ... 50th Queen's Own Regt.

1864–1865.

39.—David Kelly, 3038 ... ... 70th Regiment.

1864–1866.

40.—Jas. Carty, 484 ... ... 2nd Batt., 18th Royal
Irish Regiment.

1865–1866.

41.—Edward Elderton, 706 ... 57th Regiment.

ABYSSINIA.

General Order 18, 1st March, 1869. Her Majesty granted a
Medal to the troops who had been engaged under Lord Napier in
Abyssinia from 4th October, 1867, to 19th April, 1868.

This medal is smaller than any other issued, and is suspended
from a silver ring attached to a crown on the top of the medal.

Obverse.—Diademed head of Her Majesty, with veil, within a star ;
between the points of the star, " A B Y S S I N I A."

Reverse.—A circle, in which is stamped the recipient's name and
regiment, surrounded by laurel branches.

Riband : Scarlet, with white border.

Her Majesty's 3rd Dragoon Guards, 4th, 26th, 33rd and 45th
Regiments have " Abyssinia " on their colours.

42.—H. Barkley, 238 ... ... 1st Batt., 4th K. O. R.
Regiment.

43.—J. Conn, 312 ... . 33rd D. W. Regiment.

44.—S. Litchfield, 3633 ... ... 33rd D. W. Regiment.
Case XII., No. 15.

ASHANTEE.

General Order 48, 1st June, 1874. . Her Majesty granted a Medal to the troops engaged in Ashantee under Sir Garnet Wolseley, from the 9th June, 1873, to 4th February, 1874, and a clasp for the battle of Amoaful, inscribed " Coomassie."

Her Majesty's 23rd, 42nd, and Rifle Brigade, have " Ashantee " on their colours.

Obverse.—Diademed head of Her Majesty, with veil. " Victoria Regina."

Reverse.—Represents a skirmish between British troops and natives.

Riband : Yellow, with black border, and two narrow black stripes down the centre.

NO CLASP.

45.—T. Dawkins, Private, 1115   ...   2nd W. I. Regt., 1873-4.

CLASP—COOMASSIE.

46.—W. Clelland, Private, 130   ...   42nd Highlndrs, 1873-4.

CASE V.

CRIMEA.

" Medals of the British Army,
" Carter, Page I.

" In December, 1854, the Queen was pleased to command that a Medal, bearing the word " Crimea," with an appropriate device, should be conferred upon all the officers, non-commissioned officers, and private soldiers of Her Majesty's Army who had been engaged in the arduous and brilliant campaign in the Crimea ; and that clasps with the word " Alma " or " Inkerman " thereon were to be also awarded to such as were present in either of those battles. In February, 1855, Her Majesty granted a clasp for the action at Balaclava, and, in October following, a clasp for Sebastopol was awarded to all present between the 1st October, 1854—the day the army sat down before the place—and the 9th September, 1855, when the town was taken.

Obverse.—Head of Her Majesty (from the die of the War Medal). " Victoria Regina, 1854."

Reverse.—A figure of Victory, placing a wreath on the head of a classic warrior ; on the left, the word " Crimea."

Riband : Light blue, with yellow edges.

The clasps are ornamented with acorns.

CAVALRY.

ALMA, BALACLAVA, INKERMAN, SEBASTOPOL.

Having arranged these Medals in Régimental order, I have, for abbreviation, put A.B.I.S., to indicate the clasps attached to the Medals.

P 2

S.	1.—J. LEONARD	1st Dragoon Guards.
B.I.S.	2.—J. JAMES, Trumpeter	4th Dragoon Guards,
B.I.S.	3.—G. MOORE, Farrier ...	5th Dragoon Guards.
B.I.	4.—CHARLES PARKER, Sergt.	6th Dragoon Guards.
B.S.	5.—JOHN OVERTON ...	1st Royal Dragoons.
S.	6.—J. DAVIDSON	2nd Dragoons.
S.	7.—J. NEWTON	4th Light Dragoons.
S.	8.—JOHN MOONEY ...	6th Dragoons.
A.B.I.S.	9.—J. NYE, Sergeant ...	8th Hussars.
S.	10.—GEORGE DUNGATE ...	10th Hussars.
A.D.I.S.	11.—G. JOWETT, Sergeant	11th Hussars.
S.	12.—J. BROWN, Sergeant ...	12th Lancers.
B.S.	13.—JOSEPH BROWN ...	13th Light Dragoons.
A.B.S.	14.—THOMAS STEWART ...	17th Lancers.
B.S.	15.—T. FROST, Corporal ...	Royal Horse Artillery.
I.S.	16.—G. SYMONS, Sergeant	Royal Artillery. CASE XII., No. 31.
I.S.	17.—H. BRADLEY, Gr. & Dr.	1st Battery Royal Artill.
A.I.S.	18.—G. WILLIAMSON, Gunner and Driver ...	CASE XII., No. 33.
S.	19.—R. LEWIS, Sergeant ...	Royal Engineers. CASE XII., No. 35.
S.	20.—J. CUTHBERT, Sergeant, 875, Head Guard, Crimean Railway	Land Transport Corps.
S.	21.—HENRY TOVEY, Driver	Land Transport Corps.
B.I.S.	22.—W. McNAUGHTEN ...	Ambulance Corps.
A.B.I.S.	23.—WILLIAM WATSON ...	Acting Purveyor to the Forces, 1st Division.

INFANTRY.

A.B.I.S.	24.—R. NUTTING, 6160 ...	Grenadier Guards, 3rd Bat. CASE XII., No. 8.
A.B.I.S.	25.—J. MANN	Grenadier Guards, 3rd Bat.

A.B.I.S.	26.—R. Wells, 8688 ...	Coldstream Guards, 2nd Battalion.
A.B.I.S.	27.—J. Meredith, Private	Scots Fusilier Guards, Battalion.
S.	28.—James Conn	1st Regiment, 1st Batt.
S.	29.—William Jeffries ...	1st Regiment, 2nd Batt.
S.	80.—John Joyce	8rd Regiment, 1st Batt.
S.	81.—James Gready ...	4th Regiment, 1st Batt.
A.I.S.	82.—J. Hardman ...	7th Regiment.
A.B.I.S.	88.—T. Lewis (late 9th Foot)	9th Regiment.
S.	84.—W. Henderson, 8594	18th Regiment, P.A.L.I.
S.	85.—A. Flanery, Private	14th Regiment.
S.	86.—R. Berry ...	17th Regiment.
S.	87.—John Savage ...	18th Regiment, Royal Irish
A.S.	88.—J. Rudd ...	19th Regiment.
A.B.I.S.	89.—Michael Magraine ...	20th Regiment.
A.B.I.S.	40.—John Gaynor ...	21st Regiment.
S.	41.—Chas. Smith ...	28rd Regiment.
S.	42.—D. McIvor, 8781 ...	28th Regiment.
A.I.S.	48.—J. White	80th Regiment.
S.	44.—J. Holmes	81st Regiment.
S.	45.—H. Knowles	88rd Regiment.
S.	46.—S. Lichfield... ...	88rd Regiment. CASE XII., No. 15.
S.	47.—W. J. May	84th Regiment.
A.I.S.	48.—Thomas Walsh ...	88th Regiment.
S.	49.—J. Glover, Corp., 1681	89th Regiment.
A.S.	50.—Owen Darling ...	41st Regiment.
A.S.	51.—D. Fraser	42nd Regiment.
A.I.	52.—Michael Hannaway ...	44th Regiment.
S.	58.—A. Davey	46th Regiment.
A.I.S.	54.—John Kelly	47th Regiment.
S.	55.—Fredk. Clarke, 2468	48th Regiment.
A.I.S.	56.—J. Brown	49th Regiment.
A.I.S.	57.—Wm. Fahey, Sergeant, 8840	50th Regiment.
S.	58.—Jas. Hodges, 8776 ...	55th Regiment.

	59.—	56th Regiment.
B.I.S.	60.—J. FLANAGAN	57th Regiment.
S.	61.—J. HILLIER	62nd Regiment.
A.B.I.S.	62.—THOMAS TAYLOR, Corp., 1487	63rd Regiment.
A.B.I.S.	63.—JOHN MEARA, Corporal	68th Regiment.
S.	64.—EDWARD FARRELL, 2573	71st Regiment.
	65.—	72nd Regiment.
A.I.S.	66.—JOSEPH MARS, 1608 ...	77th Regiment.
	67.—W. CAMPBELL, Corp....	79th Regiment. CASE XII., No. 25.
A.D.S.	68.—THOMAS ROSS, 2712 ...	79th Cameron Highldrs.
S.	69.—ARTHUR GORDON, 2307	82nd Regiment.
A.I.S.	70.—P. McMAHON	88th Regiment.
S.	71.—DENS. HENNESSY, 3141	89th Regiment.
S.	72.—T. SEABRIGHT ...	90th Regiment.
S.	73.—J. MUNRO, Sergeant...	93rd Regiment. CASE XII., No. 28.
A.B.S.	74.—J. LAING, Sergeant ...	93rd Regiment.*
I.S.	75.—D. CARTHY	95th Regiment.
S.	76.—SAMUEL MILLER ...	97th Regiment.
A.B.I.S.	77.—J. DEVITT, Sergeant...	Rifle Brigade, 1st Battln.
A.I.S.	78.—JOHN COLEBORNE, Pri.	Rifle Brigade, 2nd Battln.

---

* This is the only Regiment that has Balaklava on their colours.

CASE VI.

1, 2.—LONG SERVICE AND GOOD CONDUCT.

MERITORIOUS SERVICE,

DISTINGUISHED CONDUCT IN THE FIELD.

VICTORIA CROSS.

BEST SHOT IN THE ARMY,

AND

MISCELLANEOUS MEDALS.

LONG SERVICE AND GOOD CONDUCT. (1).

This Medal was instituted by His Majesty King William IV., on the 30th July, 1830, for soldiers who had been discharged in receipt of gratuities.   To cavalry after 24 years, to infantry after 21 years.

Obverse.—A military trophy, with a shield bearing the royal arms; in the centre the arms of Hanover.

Reverse.—" For long service and good conduct."

Riband : Crimson.

1.—T. Godding, Reg. Sergt.-Major... 9th Lancers, 1887.
Case XII., No. 3.

2.—Thomas Allan, Trumpeter   ... 2nd Reg. Drgns., 1833.

3.—J. Greenwood, Colour Sergeant 51st Regiment.
Case XII., No. 19.

4.—H. A. Gilbert, Sergeant-Major Sappers and Miners.
Case XII., No. 36

LONG SERVICE AND GOOD CONDUCT (2).

When Her Majesty succeeded to the throne, a new Medal was struck without the arms of Hanover, in consequence of that kingdom having passed from this country.   The Medal, riband, etc., are otherwise the same.

On the 16th January, 1860, by Royal Warrant, the medal was granted to such soldiers as had fulfilled the required conditions without gratuities, in consequence of the sum granted having been exceeded.

5.—Robert Nutting, Private, 6169 1st Batt. Gren. Guards.
Case XII., No. 8.

6.—Wm. Rofe, Corporal   ...   ... 2nd Regiment.
Case XII., No. 9.

7.—Edward Taylor, Sergeant, 2543 91st Foot.

MERITORIOUS SERVICE.

Authorised by Royal Warrant, 19th December, 1845. That a sum not exceeding £2,000 a-year should be distributed in granting annuities as rewards for distinguished or meritorious service, to sergeants recommended by the Commander-in-Chief, in sums not exceeding £20 per annum.

By Royal Warrant, 4th June, 1853, the sum was increased to £4,000.

By Royal Warrant, 4th December, 1854 :—

" To mark Her Majesty's sense of the gallantry of the Army in the Crimea, one sergeant of each Cavalry, Infantry, and battalion of Foot Guards should be selected, provided the grants did not exceed £4,000 in any one year."

Sergeants of the Royal Marines also received this medal.

Obverse.—Head of Her Majesty. "Victoria Regina."

Reverse.—Two laurel branches. "For Meritorious Service;" surmounted by a crown.

Riband : Crimson.

8.—MAURICE REARDON, Sergnt.-Major   102nd Foot.

DISTINGUISHED CONDUCT IN THE FIELD.

Authorised by Royal Warrant, 4th December, 1854.

" For marking the Sovereign's sense of the distinguished service and gallant conduct in the field of the Army then serving in the Crimea, under Field-Marshal Lord Raglan."

Obverse.—A military trophy, same as the Long Service Medal.

Reverse.—" For distinguished conduct in the field."

Riband : Crimson with dark blue stripe down the centre.

9.—G. WILLIAMSON, Gunner and Driver ...   Royal Artillery.
CASE XII., No. 38.
10.—WM. FAHEY   ...   .   50th Regiment.
11.—CHARLES JOHNSON   ...   .   77th Regiment.

VICTORIA CROSS.

Was instituted by Her Majesty by Royal Warrant, 29th June, 1856.

" Whereas we take into our Royal consideration that there exists no means of adequately rewarding the individual gallant services either of officers of the lower grades in our naval military service, or of warrant and petty officers, seamen and marines in our navy, and non-commissioned officers and soldiers in our army."

There are fifteen clauses relative to this decoration in the Warrant.

5th Clause.—" It is ordained that the cross shall *only* be awarded to those officers or men who served us in the presence of the enemy, and shall have then performed some signal act of valour or devotion to their country."

Obverse.—Bronze Maltese cross, with Royal crest in centre. " For valour," on a scroll."

Reverse.—The date of the act of valour.

The cross is worn suspended by a V from a clasp on which is engraved the rank, name, and regiment of recipient.

Riband : Crimson.

V.C.   12.—Charles Anderson   ...   2nd Dragoon Guards.
            8th October, 1858.

For an act of gallantry in saving the life of Colonel Seymour, C.B., during the Indian mutiny.

V.C.   13.—Geo. Symons, Sergeant   Royal Artillery.
            6th June, 1855.

Class XII., 13.

For conspicuous gallantry in having volunteered to unmask the embrasures of a five-gun battery in the advanced right attack, and when so employed under a terrific fire which the enemy commenced immediately on the opening of the first embrasure, and increased on the unmasking of each additional one ; in having overcome the great difficulty of uncovering the last by boldly mounting the parapet and throwing down the sandbags, when a shell from the enemy burst and wounded him severely.

V.C.   14.—T. FREEMAN, Private ... 9th Lancers.
       10th October, 1857.

For having gone on the 10th October, 1857, at Agra, to the assistance of Lieutenant Jones, who had been shot; killing the leader of the enemy's cavalry, and defending this officer against several of the enemy.

V.C.   15.—JOHN KIRK, Private ... 10th Regiment.
       4th June, 1857.

For daring gallantry at Benares on the 4th June, 1857, on the outbreak of the mutiny of the native troops at that station, in having volunteered to proceed with two non-commissioned officers to rescue Captain Brown, Pension Paymaster, and his family, who were surrounded by rebels in the compound of their house, and having at the risk of his own life succeeded in saving them.

V.C.   16.—JAMES MUNRO, Colour-
       Sergeant   ... 93rd Highlanders.
       16th November, 1857.

Case XII., No. 28.

Extract from *London Gazette*, 8th November, 1860 :—

" For devoted gallantry at Secunder-Bagh on the 16th November, 1857, in having promptly rushed to the rescue of Captain E. Walsh of the same corps when wounded and in danger of his life, whom he carried to a place of comparative safety, to which the sergeant was brought in very shortly afterwards badly wounded."

He received the cross from Her Majesty's hands at Windsor Castle, 9th November, 1860.

BEST SHOT IN THE ARMY.

Obverse.—Head of Her Majesty, diadem and veil. " Victoria Regina."

Reverse.—A figure of Victory rising from a throne, in the act of crowning an ancient warrior armed with shield and bow and arrows. In her left hand a horn.

Riband : Red, with two black narrow stripes, with a white one between them on each edge.

17.—This medal is given annually to the best shot of the year in usual musketry practice, with a gratuity of £20.

THE ORDER OF THE BATH.

K.C.B.

The badge of a Knight-Commander of the Order is a gold Maltese cross of eight points enamelled, terminating in small gold balls, having in each of the four angles a gold lion of England. In the centre, on a ground of white enamel, are the rose, thistle, and shamrock, issuing from a gold sceptre between three gold Imperial crowns, all within a red circle charged with the motto, " Tria juncta in uno," in gold letters, surrounded by two branches of laurel in the proper colours, issuing from an escrol of blue enamel, containing the words " Ich Dien " in letters of gold.   The cross is 3¼ inches in depth and width.

The badge of K.C.B. is attached to the riband by a large gold ring chased with oak leaves and acorns, and is suspended from the neck.

Riband : Crimson, four inches.

The star is of silver in the form of a cross patée, having three Imperial crowns in the centre, surrounded by the motto and branches of laurel with the escrol, " Ich Dien," and is worn on the left breast.

18.—These belonged to Lieutenant-General Sir C. F. Smith, R.E., K.C.B., who was nominated 23rd September, 1848, and was invested by dispensation 12th June, 1814.

C.B.

19.—Badge similar to the K.C.B., only smaller, and is fixed to a gold swivel, with a bar of the width of the riband, two inches, and fastened to the coat by a gold buckle.

Although this most ancient order of knighthood dates from the earliest days of English history, it bore but little resemblance to the military Order of the Bath as created by King George I., 1725, and enlarged and re-organised in 1815.

———————

Miscellaneous Medals.

IN CASE VI.

CULLODEN.

GOLD.

Obverse.—Head of the Duke of Cumberland. " Cumberland."
Reverse.—Figure of Apollo. Dragon pierced by an arrow.
" Actum est ilicet Periit. Prœl: Colod: Ap. XVI., MDCCXLVI."

This Medal was given to officers who commanded regiments at the battle of Culloden; and was worn round the neck by a crimson riband with a green border.

1.—This Medal belonged to Brigadier-General Fleming. He commanded the 36th Regiment at the battle of Culloden.

BADGES OF CHARLES I.

There were several badges bearing the head of Charles I., and also of Queen Henrietta Maria; but whether they were ever worn is doubtful.

1.—Supposed for the Battle of Naseby.

Obverse.—Head of General Fairfax.
Reverse.—" Post Hac Meliora. Meruisti 1645."
2.—Head of Charles, etc.
3.—   ,,   ,,

4.—This Medal was supposed to have been given after the Battle of Dunbar.

Obverse.—Head of Cromwell.   Word at Dunbar, " The Lord o Hosts, Septem. 18, 1650."
Reverse.—The House of Commons sitting.

5.—Head of General Monk, Duke of Albemarle.

Reverse.—Coat of arms.

ROYAL AFRICAN CORPS.

A LARGE SILVER MEDAL.

Obverse.—Bust of King George III., in mantle and collar of the Garter.   " Georgius III., Dei Gratiâ, Britanniarum Rex, F.D."
Reverse.—Royal Arms, 1814.

On this Medal is inscribed, " The Gift of Colonel McGartey, Governor of the West Coast of Africa, to Sergeant John Harris, of the Royal African Corps, for his valiant behaviour in capturing a Spanish brig of 18 guns, with 1 gun and 11 men in a gunboat. This victory was achieved on the 7th April, 1818.

Riband : Crimson, with blue edge.

CAPE CORPS AND NATIVE LEVIES.

A LARGE MEDAL.

Obverse.—A lion rampant, with laurel wreath, 1851.
Reverse.—" Presented by His Excellency Sir H. G. Smith, Bart., G.C.B., to H. McKain, for gallantry in the field."
Riband : Dark blue, red edge.

The following extract from a letter from Sir H. Smith was sent to me by Sir E. Holdich, K.C.B., who was Aide-de-Camp to Sir H. Smith :—

"From Sir H. Smith to Mr. Montague, Colonial Secretary.

" King William Town,
" 7th March, 1851.

" I wish you would have some 30 or 40 Medals struck, made in same way of the make of the Waterloo Medal, but about two-thirds of the size, with same device, if we can make it. I wish to give them to some of the Cape Corps, and of the Levies who distinguished themselves.

Inscription : " Presented by Sir H. Smith, for gallantry in the field."

YEOMANRY.

Obverse.—" West Somerset Yeomanry." A mounted Hussar. " C.K.K. Tynte, Col."

Reverse.—" Merit." Troop Sergeant-Major John Carter, Bridgewater Troop, May, 1835.

Riband : Blue.

UNKNOWN.

AN ENGRAVED MEDAL.

Obverse.—An elephant, with castle, etc. ; " Hindostan ; " laurel wreath.

Reverse.—" For services in India."

A STAR OF SEVEN POINTS (BLACK CENTRE).

96th.—" Veni, vidi, vici."

Supposed to be a Temperance Medal of the 96th Regiment.

CASE VII.

REGIMENTAL MEDALS

AND

DECORATIONS.

These medals were given to the officers of the different regiments, to their men, either to commemorate some particular act of bravery, or for services in the field, or for long service and good conduct, and were discontinued when the long service and good conduct medal was instituted in 1837.

9TH (THE QUEEN'S ROYAL) LANCERS.

1.—*Obverse.*—Two lances crossed; in the centre, 9. Above, a crown; below, the regimental cypher. Round the medal, " Queen's Royal Lancers," " Peninsula."

Reverse.—" Presented to Regimental Sergeant-Major Thomas Godding, by the officers of his Regiment, as a token of esteem and in testimony of his faithful and meritorious services for upwards of 32 years, March, 1837." Round the above, " Present at the sieges of Buenos Ayres and Flushing; at the Battles of Arroyo de Molinos, Vittoria, St. Sebastian, Pyrenees, Orthes, and Toulouse.

CASE XII., No. 3.

(2) A SMALLER MEDAL.

2.—*Obverse.*—The same.

Reverse.—" Edinburgh, 20th March, 1837. Presented to Regimental Sergeant-Major Thomas Godding, by the non-commissioned officers of the Regiment, as a token of esteem and in testimony of his gallant and meritorious services for upwards of 32 years."

CASE XII., No. 3.

Riband: Crimson, blue edge.

10TH (THE PRINCE OF WALES' OWN) ROYAL HUSSARS.

3.—*Obverse*—Prince of Wales Plume, surrounded by a garter,

bearing the words "Prince of Wales' Own ;" below, on a scroll, "Royal Hussars." The whole surrounded by wreaths of roses, shamrock and thistles.

Reverse.—A mounted Hussar, 1843 ; surrounded by a similar wreath.

Round the edge : Presented to Sergeant John Day, by his brother non-commissioned officers, 10th Royal Hussars, as a mark of esteem.

Riband : Dark blue.

14TH (THE KING'S) LIGHT DRAGOONS.

4.—*Obverse*—Within a wreath, " Fortitudine. Blascho Sancho, 26th July, 1812. Peninsula."

Reverse.—Within a wreath, " William Hanley, Corporal, 14th Light Dragoons."

CASE XII., No. 5.

Riband : Crimson.

Sergeant William Hanley, in a letter addressed to the Editor of the *U.S. Journal*, November, 1840, page 382, 384, relates his capture of a French picquet at Blascho Sancho, on the 26th July, 1812 ; as an honourable testimony of which, he was presented with this medal at the head of his regiment. *Cannon's Historical Records*, page 39, 40, thus details the event :—" On the following day the regiment pursued the rear of the French army, and two squadrons were sharply engaged, and took several prisoners near Tenerada. On the 26th, a patrol of three dragoons of the 14th, and and four of the German Hussars, under Corporal William Hanley, of the former corps, detached to Blascho Sancho, captured a party of the enemy, consisting of two officers, one sergeant, one corporal, and 27 mounted dragoons, with one private servant and two mules, for which they received the expressions of the approbation of the commander of the forces. The French horses were given to the 14th and German Hussars, to complete deficiencies. The amount was divided among the patrol, and a further pecuniary donation was afterwards made to the men engaged in this gallant exploit."

16TH (THE QUEEN'S) LIGHT DRAGOON LANCERS.

5.—A large Maltese cross inscribed, " A gift from Lieutenant-General Sir John Vandeleur to Francis Lambert, of the 16th (the Queen's) Light Dragoon Lancers. Oporto, Salamanca, Vittoria, Nive, Peninsula, Talavera, Fuentes D'Onor, Busaco."

Suspended from a ring, with clasp ; surmounted by a royal crown.
Riband : Light blue.

22ND LIGHT DRAGOONS.

6.—*Obverse.*—A crown above a rose, shamrock, and thistle. " Reward of Merit and Faithful Service."

Reverse.—XXII $\frac{L}{D}$; within a laurel wreath, 1815. " Seringapatam, 1799 ; Conicul, 1800 ; Java, 1811."

Riband : Light yellow.

1ST (THE ROYAL) REGIMENT OF FOOT.

7.—*Obverse.*—" Presented by Lieutenant-Colonel George Bell and the Officers of the 2nd Battalion Royal Regiment."

Reverse.—" To Drum-Major George Morgason, for long service and exemplary conduct."

Suspended from a clasp, with " 36 years' service—never a defaulter " on reverse of clasp. " 1847 " inscribed on it.

Riband : Red.

2nd (THE QUEEN'S ROYAL) REGIMENT OF FOOT.

A Maltese cross, with the Paschal lamb in bold relief in the centre ; on the arms of the cross, " Queen's Royal—Merit," and a Roman numeral, with the word " years," denoting the length of service of the recipient.

Reverse.—Queen's Royals. For ten years' meritorious service to—

    8.—JAMES WILKINS.

    9.—WILLIAM ROFE.

CASE XII., No. 9.

For six years, to—

    10.—EVAN DAVIES.

This Medal is suspended from a clasp, with the words, " For meritorious and good conduct."

Riband : Dark blue.

The distinction between the two grades is—for ten years, a silver cross, with a gilt lamb ; for six years, a bronze cross, with a silver lamb.

The crosses were given, with a certificate, as follows :—

" James Wilkins, of the Queen's Royals, having established his claim to the honourable distinction of a Silver Regimental Cross for ten years' meritorious conduct in the corps, and being recommended by the Board which assembled on the 29th January, 1834, for the same, it is herewith presented to him, with this certificate from the commanding officer.

" Sealed and dated at the Camp, near Poonah, 8th February, 1834.

    (Signed)   " THOMAS WILLSHIRE, Colonel.

        " Lieut.-Col. Commanding Queen's Royals."

5TH NORTHUMBERLAND FUSILIERS.

11.—(1.) *Obverse.*—St. George and the Dragon ; on one side, Vth, on the other, F$^{t.}$ " Quo fata vocant."

Reverse.—" Reward of 14 years' military merit, 18th January, 1769," within a laurel wreath.

This Medal is a curiously-engraved one, with a loop for suspension.

12.—(2.) *Obverse.*—St. George and Dragon. " Quo fata vocant," on scroll. " Revived April 23rd, 1805."

Reverse.—Vth Foot. Merit. March 10th, 1767; with laurel leaves.

13.—(3.) The same, with the exception of " Revived April 23rd, 1805," on the obverse.

14.—(4.) *Obverse.*—The same as No. 3.

Reverse.—" Northumberland Fusiliers. Merit. March 10th, 1767; " within a laurel wreath.

15.—(5.) The same as No. 4, in bronze.

Riband : Dark green.

Early in the year 1767, a system of honorary distinctions for long continued good behaviour was introduced into this Regiment, which was formed to stimulate the indifferent to good conduct, and those already worthy to perseverance in well doing, and it produced such a body of non-commissioned officers as few corps could boast of.

These distinctions were of three classes, to be worn suspended by a riband at a button hole of the left lappel. The first or lowest class, which was bestowed on such as had served irreproachably for seven years, was of bronze, bearing on one side the badge of the regiment, St. George and the Dragon, with the motto, " Quo fata vocant," and on the reverse, " Vth Foot. Merit." The second class was of silver, and the third was similar to the second, with the recipient's name engraved on it. For twenty-one years' good and faithful service a soldier had received from his commanding officer this honourable testimony of his merit. These medals were bestowed only on soldiers who, for their respective periods of seven, fourteen, or twenty-one years, had never been tried by Court Martial. They were given by the commanding officer on parade, and were as publicly cut from the soldier s breast by the drum major, if he committed any crime through which he forfeited it. Those who obtained the third class had also an oval badge of the colour of the facing, on the right breast, embroidered with gold and silver wreaths, and the word " Merit " in gold letters.

On the 7th November, 1768, Lieutenant-General Hodgson was succeeded in command of the 5th by Hugh Earl Percy, afterwards the Duke of Northumberland, who duly estimated the value of this regimental " order of merit," and kept it up with all the liberality and dignity it deserved, and the following order issued by him on the subject is referred to in Adye's "Essay on Rewards and Punishments," viz :—

" Earl Percy, having perceived with great pleasure the happy effects of the regimental ' Medal of Merit,' influencing the non-commissioned officers and soldiers of the Fifth to deserve the favour of their officers ; and being anxious, as far as may be in his power, to encourage them to persevere in such sentiments of honour, is determined, for the future, to give them out every year a short time before the Review, instead of the usual day, as it often has happened that the regiment has been separated, which prevented the men who were entitled to that mark of honour from receiving it in so public a manner as his lordship should wish."

Previous to the embarkation of the 5th from Ireland, the circumstance of the regiment having an " Order of Merit " attracted the attention of the local military authorities. The commanding officer, after the arrival at Gibraltar, was, in consequence, called upon by the General Commanding-in-Chief, Lord Hill, to explain under what regulations and arrangements the Order was conferred, and other particulars of the institution. The required information was promptly afforded, and elicited the following reply :—

" Horse Guards,

" 20th June, 1832.

" Sir,

" I have had the honour to submit to the General Commanding-in-Chief your letter of the 4th instant, with its enclosures on the subject of the ' Order of Merit ' existing in the Fifth Foot, and am directed to acquaint you that the explanation afforded by Lieutenant-Colonel Sutherland shows that the Order in question is dispensed under the most laudable regulations, and has been productive of the best effects during the long period since its original establishment in the regiment. It is considered highly desirable, however, that both officer and soldier should be taught, under all circumstances, to expect professional honours from the Sovereign alone ; and under this impression Lord Hill has been induced to recommend to the King to give the royal authority for the confirmation and continuance of this regimental badge of distinction—an arrangement which, while it bestows upon it legitimate existence, will at the same time no doubt enhance its value in the estimation of those on whom it is conferred. You will therefore be pleased to communicate this decision to Lieutenant-Colonel Sutherland, and acquaint him that he is at liberty to proceed in the distribution of the medals and badges as heretofore.

" I have, etc.,

(Signed) " JOHN MACDONALD, Adj.-Gen.

" To Lieutenant-General Sir William Houston, G.C.B., G.C.H.,

" Commanding at Gibraltar, 1832."

Cannon's Historical Records, pp. 87, 89, 95, 97.

———

7th ROYAL FUSILIERS.

SILVER.

16.—*Obverse.*—Pallas and a figure of Victory crowning a warrior, seated, with a laurel wreath. " Order of Merit, established MDCCLXXXVIII."

Reverse.—A full-blown rose surrounded by garter, " Honi soit qui mal y pense," surmounted by a Royal crown ; the whole radiated.
" Military virtue rewarded. VIIth Regiment or Royal Fusiliers."

BRONZE.

17.—*Obverse.*—Rose, etc., as above ; surrounded with the words " Order of Merit, established MDDCLXXXVIII."

Reverse.—The same as the silver medal.
Riband ; Dark blue.

13th, OR PRINCE ALBERT'S, REGIMENT OF LIGHT INFANTRY.

GOLD.

17A.—*Obverse.*—A bugle on which are the words " Ava," " Martinique ;" within the cords of the bugle, XIII. ; above, a sphinx with the word " Egypt." Round the whole, " Medal of Merit for 20 years' good conduct."
Reverse.—Quite plain.
This medal belonged to William Webber.

SILVER.

The same as above.
18.—(1.) " For 14 years' good conduct."
19.—(2.) " For 7 years' good conduct." Private J. ROUND.
CASH XII., No. 11.

Riband ; Yellow, and red edges.

16th FOOT.

20.—*Obverse.*—" Sobriety," surrounded by a wreath of roses.
" Temperance Society, 16th Foot."
Reverse.—Awarded to I. Gannon, 1838.
Riband : Yellow.

20th (THE EAST DEVONSHIRE) REGIMENT OF FOOT.

GOLD.

21.—*Obverse.*—A crown, xx, and a sphinx ; in the background two flags crossed ; the words " Omnia audax," surrounded by laurel branches.

Reverse.—xx in the centre, and radiating it the following : " For highly meritorious conduct during a period of 18 years, June 18th, 1838. Minden, Egmont-op-Zee, Egypt, Maida, Vittoria, Pyrenees, Orthes, Toulouse, Peninsula, Vimiera, Corunna," surrounded by laurel branches.

Round the edge,—Presented by his brethren in arms XXth Regiment to John Dorrington, Quartermaster-Sergeant.

Riband : Yellow.

22nd (THE CHESHIRE) REGIMENT OF FOOT.

GILT.

22.—*Obverse*—Hercules crowning a warrior with a wreath; in the background a camp. " Order of Merit, estab : MDCCLXXXV."

Reverse.—With a wreath. " Reward for military virtue, from Lt.-Col. Crosbie ; " outside the wreath, " xxii or Cheshire Regiment."
Riband : Blue.

This was one of the original medals, which was discontinued, subsequently replaced by the following :—

SILVER.

23.—*Obverse.*—A soldier kneeling, receiving a medal from the King on the terrace at Windsor Castle. " Established under Royal sanction, 1785."

Reverse.—" Order of Merit, 22nd Regiment." " Re-established by Col. Sir H. Gough, 1st January, 1820.

J. Connolly, Sergeant.

BRONZE.

24.—Same as silver.

Riband : Some have red, blue, or yellow.

In 1785, when the regiment was stationed at Windsor, under the command of Lieutenant-Colonel Crosbie, an order of merit was instituted in the corps with the view of promoting good order and discipline. The field officers, captains, and adjutant for the time being, to be members of the order. The order consisted of three classes. The first wore a silver-gilt medal, suspended to a blue riband two inches broad, and worn round the neck. The second, a silver medal ; and the third, a bronze, similarly worn. The candidates for the third class must have served seven years ; for the second, fourteen years ; and for the first, twenty-one years, with an unblemished character.

On July 1st, the King was graciously pleased to accept by Colonel Crosbie, a medal of the first class, and on the 3rd, the regiment being encamped in Windsor forest, assembled a parade, with the non-commissioned officers and soldiers selected to receive the medal in front. The rules of the order were read ; the men presented arms ; and the band played " God save the King."

Cannon's Hist. Records, p. 15.

Extract from a letter from the late Sir John Pennefather :—

" All the information in my possession respecting the 22nd regimental medal is, that I believe it was instituted by Colonel

Crosbie, who commanded the regiment in 1785, then quartered at Windsor; and that it was so far legalized that King George III. received one of the medals on a parade of the regiment, from Colonel Crosbie. The regiment went to the Cape of Good Hope in the latter years of the last century; thence to India. The command fell into various hands; the old soldiers died off; and when the regiment came home from the East in 1819, the medal had got into desuetude. When the regiment came home in 1819, Sir Hugh (now Lord) Gough was appointed to its command; and, I suppose, reading about Colonel Crosbie's medal, re-established the order in his own name, with a new die, and under altered regulations. Sir Hugh received orders to discontinue the wearing of the medal some years after (he left the 22nd in 1826) from the Horse Guards. Sir Henry Torrens being then, as well as I can recollect, Adjutant-General. That is all I know about it.

" August 24th, 1864."

23RD ROYAL WELSH FUSILIERS.

25.—*Obverse*—Prince of Wales' plume and motto, " Ich Dien ;" above, the word " Wellington ;" below, " Peninsula," surrounded by laurel branches.

Reverse.—" Albuhera, Badajoz, Salamanca, Vittoria, Pyrenees, Nivelle, Orthes, Toulouse," surrounded by wreaths.

Round the edge, " Evan Williams, 23rd R.W.F."

Riband : Crimson, blue edge.

26TH (THE CAMERONIAN) REGIMENT OF FOOT.

Obverse.—A female figure, seated with an open book in her lap, presenting the medal to a soldier; the emblems of faith and hope at her side.   In the background a camp; above, " Thou art worthy;" below, " Be thou faithful unto death."

Reverse.—The Regimental Colours crossed over a Bible ; above, a thistle and crown; the whole encircled by wreaths. " 1828," below.

Riband :

    26.—Wm. Galloway, Qrmstr. Sergt.   26th Regiment, 1827.
    27.—E. Kenny, Sergeant ...   ...   26th Foot.
                                       Case XII., No. 13.

    28.—(2.) *Obverse.*—" Merit," within a wreath of thistles ; " Cameronian Regimental School,"

Reverse.—Quite plain.

Riband : Yellow.

34th (CUMBERLAND) REGIMENT OF FOOT.

    29.—*Obverse.*—Head of Duke of Wellington. " Wellington."

Reverse.—" Presented by Lieut.-Colonel Airey and the officers of the 34th Regiment, to Barrack-Sergeant Moses Simpson, of the 2nd Batt. 84th Regt., in commemoration of his gallant conduct as Sergeant of Grenadiers at the action of Arroyo de Molinos, in Spain, on the 28th October, 1811, when he took from the Drum Major of the French 84th Regt. of the line, the Regimental Staff, which has ever since been carried at the head of the British 34th Regiment, June, 1848."

Riband : Crimson, blue edge.

Cannon's Historical Records, p. 52.

37th (THE NORTH HAMPSHIRE) REGIMENT OF FOOT.

    30.—*Obverse.*—" Presented by Sir Alexander Duff, Colonel of the

87th Regiment, to John Howard, Bugler, in testimony of long and meritorious service for upwards of 30 years in that corps."

Reverse.—The arms of the Duff family. "Deus juvit Deo juvante," on a scroll.

Riband : Crimson, with blue edge.

40TH (THE 2ND SOMERSETSHIRE) REGIMENT OF FOOT.

31.—*Obverse.*—"40th Regiment. German Town, October 4th, 1777," surrounded by laurel wreaths.

Reverse.—The American troops investing the Store House (Chew's House), which the regiment held successfully against them.

Riband : Narrow dark-blue string.

42ND ROYAL HIGHLANDERS (THE BLACK WATCH).

32.—*Obverse.*—Troops defiling through a mountainous country. " Pyrenees " above a figure of St. Andrew, with the motto, " Nemo me impune lacessit," with a thistle on each side.

Reverse.—" Corunna, Fuentes D'Onor, Pyrenees, Nivelle, Nive, Orthes, Toulouse, Peninsula ; " above, a flying figure of Victory ; below, " 42nd Regiment," surrounded by laurel wreaths.

Riband : Dark blue.

A. MUNRO, Sergeant.

33.—(2.) *Obverse.*—" Nive, Orthes, Toulouse, Waterloo."

Reverse.—" XLII " within a laurel wreath, suspended from a clasp with " Victoria " engraved on it.

Riband : Crimson, with blue edge.

This medal was given to me by Lieutenant-Colonel J. Johnstone, whose father, Colonel George Johnstone, commanded the 42nd from 1839 to 1843.

48TH (THE NORTHAMPTONSHIRE) REGIMENT OF FOOT.

34.—*Obverse.*—48 below a crown; on a scroll, the recipient's name, Wm. Travis, 1819. "Northamptonshire."

Reverse.—" Talavera, Albuhera, Rodrigo, Badajoz, Salamanca, Vittoria, Pyrenees, Nivelle, Orthes, Toulouse," surrounded by a wreath.

Riband : Crimson, blue edge.

52ND (OXFORDSHIRE) LIGHT INFANTRY.

35.—*Obverse.*—A bugle, LII., 6th April, 1812, within a laurel wreath.

Reverse.—" A volunteer in the leading column of attack at the assault on Badajoz. W. Ben. Housley, 52nd Regiment. The officers, 52nd Regiment."

CASE XII., No. 17.

Riband : Crimson, blue edge.

53RD (THE SHROPSHIRE) REGIMENT OF FOOT.

36.—A silver plate, with " 2nd Batt. 53rd Regt., or Shropshire," within a wreath, with a clasp inscribed " Salamanca."

Sergeant Cox.

Riband : Blue and red, in two equal stripes.

These clasps were given by Colonel Sir George Bingham, compliance with instructions from the Colonel of the Regiment, Lieutenant-General Sir John Abercrombie, G.C.B. Fifteen Sergeants only got this decoration. Sergeant Cox, prior to volunteering into the 53rd, served in the 9th Regiment during the Rebellion in Ireland. He went to Oporto in 1809, and served throughout the whole Peninsula War. He had the War Medal for Talavera and Salamanca.—*Cannon's Historical Records*, p. 55, 56.

71st (HIGHLAND) LIGHT INFANTRY.

37.—(1.) *Obverse*—" Hindoostan ; " a crown ; " 71 Highland Light Infantry," on a scroll ; below, a bugle and " 30 years service. For courage, good conduct, and faithful service," on a scroll. Rose, shamrock, and thistle.

Reverse.—Figure of St. Andrew ; motto, " Nemo me impune lacessit." " Peninsula, Roleia, Vimiera, Almaraz, Vittoria, Nive, Pyrenees, Orthes, Waterloo, Fuentes D'Onor."

Riband : Crimson, blue edge.

ROBERT MUNRO, Colour-Serjeant, May, 1834.

38.—(2.) *Obverse.*—" For courage, loyalty, and good conduct;" on a scroll ; a crown, " 71," and " 21 years' service."

Reverse.—" Tria juncta in uno," on a scroll ; rose, shamrock, and thistle.

Riband : Same.

THOMAS ANDERSON, Sergeant.

39.—(3.) Shooting badge. A bugle in the centre of a St. Andrew's cross ; " 71 ;" suspended from a clasp, on which is " Prize shot, 1840."

Riband : Yellow.

74TH (HIGHLAND) REGIMENT OF FOOT.

SECOND CLASS.

40.—(1.) *Obverse.* " 74th," within a wreath.   " Rodrigo, Badajoz, Fuentes."

Reverse.—The same, with " Vittoria, Orthes, Toulouse.'

Riband : Yellow.

H. CRABTEE.

THIRD CLASS (Smaller Size).

41.—(2.) *Obverse.*—Same, with " C. Rodrigo, Pyrenees."

Reverse.—Same, with " Orthes, Toulouse."

In consideration of the meritorious conduct of the non-commissioned officers and men of the Regiment during the Peninsula war, Colonel French applied to the Commander-in-Chief to authorize those most distinguished amongst them to wear medals in commemoration of their services.   The sanction of the Commander-in-Chief was conveyed to Colonel French in a letter from the Adjutant-General, dated " Horse Guards, June 30th, 1814."

Medals were accordingly granted, which were divided into three classes.

First Class : For those who had served in eight actions and upwards.

Second Class : For six or seven actions.

Third Class : For four or five actions.

Cannon's Hist. Records, p. 103.

77TH (EAST MIDDLESEX) REGIMENT OF FOOT.

——

42.—*Obverse.*—Prince of Wales' plume, with crown and motto, "Ich Dien;" below, "77." "Peninsula," on a scroll, joining two laurel branches.

Reverse.—"Elboden, Ciudad Rodrigo, Badajoz," encircled with laurel branches.

Riband : Crimson, blue edge.

DAVID NARGATE, Drummer.

——

79th (CAMERON HIGHLANDERS) REGIMENT OF FOOT.

——

48.—*Obverse.*—A military trophy in the centre, a shield with "79" inscribed.

Reverse.—A radiated star ; in the centre, a sphinx, with "Egypt," "Waterloo and Peninsula." Round the star, scrolls with "Egmont-op-Zee, Fuentes D'Onor, Salamanca, Pyrenees, Nivelle, Nive, Toulouse."

Riband : Dark green.

H. BANNERMAN, Sergeant.
GEORGE McKAY, Sergeant.

Sergeant G. McKay had also a medal almost similar to the 42nd Regiment medal, and evidently altered from it, with a clasp inscribed "George McKay, 79th Regiment."

This medal was instituted in 1819, by the late Lieutenant-General Sir Neil Douglas, K.C.B., who served with the regiment through the Peninsula War, and commanded it at Waterloo. There were two classes—a bronze medal for men with seven years service, who had not been tried by court martial; a silver medal after twelve years, when the bronze medal was returned and given to the next entitled. They were discontinued in Dublin, in 1888, but the men in possession of them were allowed to wear them as long as they served.

88th REGIMENT OF FOOT, CONNAUGHT RANGERS.

FIRST CLASS.

(1.)—A Silver Maltese cross, on the arms of which are inscribed : " Talavera, Busaco, Nive, Fuentes D'Onor, Pyrenees, Nivelle, Ciudad Rodrigo, Toulouse, Orthes, Salamanca, Vittoria, Badajoz." In the centre, " 88."

Reverse.—The recipient's name.

44.—Thos. Kelly, Sergeant.
45.—Wm. Kavanagh, Sergeant.

Sergeant Kavanagh was transferred from the 88th to the 45th Regiment ; during his services in the 88th he gained the 1st class or regimental cross. On joining a regiment that had no regimental decoration, and unwilling to appear without his 88th cross, he had the number of his new regiment fixed over the 88 ; which number came off in cleaning it, and revealed the 88.

Verified at the Horse Guards.

SECOND CLASS.

Obverse.—A female figure seated, holding in her outstretched right hand a wreath ; her left hand supports an Irish harp.

Reverse.—" 88," and, within a laurel wreath, the actions for which the recipient received the medal.

46.—(1.)—" Orthes, Busaco, Badajoz, Vittoria, Salamanca, Fuentes D'Onor, Ciudad Rodrigo, Pyrenees, Toulouse, Nivelle, Nive."

(Eleven actions.)
James Tracey.

47.—(2.)—" Orthes, Busaco, Badajoz, Vittoria, Salamanca Fuentes D'Onor, Ciudad Rodrigo, Talavera, Toulouse, Nivelle, Nive."

(Eleven actions.)

JOHN CUNNINGHAM, Corporal.

48.—(3.)—" Orthes, Badajoz, Busaco, Salamanca, Fuentes D'Onor, Ciudad Rodrigo, Pyrenees, Toulouse, Nivelle, Nive."

(Ten actions.)

WILLIAM IRWINE, Corporal.

49.—(4.)—" Vittoria, Orthes, Busaco, Toulouse, Fuentes D'Onor, Pyrenees, Nivelle, Nive, Badajoz."

(Nine actions.)

EDWARD CONNAH.

50.—(5)—" Orthes, Badajoz, Vittoria, Pyrenees, Fuentes D'Onor, Toulouse, Nivelle, Nive."

(Eight actions.)

———

THIRD CLASS.

51.—(1.)—" Nive, Vittoria, Pyrenees, Toulouse, Nivelle, Orthes."

(Six actions.)

JAMES KELLY.

52.—(2.)—" Orthes, Vittoria, Pyrenees, Fuentes D'Onor, Toulouse, Toulouse, Nivelle."

(Six actions.)

53.—(3.)—" Toulouse."

Riband : Was crimson with blue edge.

When quartered in Edinburgh in 1818, it occurred to Colonel Wallace to establish an Order of Merit in this Regiment. Colonel Wallace applied to the Commander-in-Chief for leave to grant this medal.

Edinburgh Castle, June 4th, 1818.

" Sir,

" I have the honour to state that some of the non-commissioned officers and soldiers of the Eighty-eighth Regiment have served in twelve different general actions, and have been two, and three, and four times wounded ; have been a long time in the regiment and always conducted themselves well in the field and in quarters. I am anxious to bestow upon them some mark of distinction for this good conduct, as an encouragement to them and others in future. I shall be much obliged to you if you will obtain His Royal Highness the Commander-in-Chief's permission for me to give such men medals as a testimony of their merit.

"I have, etc.,

"J. A. WALLACE,

"Colonel Commanding.

" To the Adjutant General."

The answer of the Commander-in-Chief, communicated in a letter from Sir Henry Torrens, dated 28th of the same month, sanctioned Colonel Wallace's plan, leaving it to his discretion to grant such testimonials in the case alluded to, as he might deem essential to the good of the regiment.   The proper authority thus obtained, Colonel Wallace's intentions were carried into effect without delay. Silver Medals of three classes were struck, at the expense of the officers of the regiment.

The first class was given to men who had been in twelve general actions.   The second to those who had been in from six to eleven. The third to those who had been in six, or any less number.

The Medal was worn from a bar inscribed " Peninsula."

The numbers distributed were :—

	Sergts.	Corps.	Drum.	Privates.
1st Class	13	6	6	45
2nd Class	7	9	3	126
3rd Class	19	10	3	185
	39	25	12	356

94TH REGIMENT OF FOOT.

54.—*Obverse.*—An elephant and crown, xcɪv. ; " Scotch Brigade." Bunches of thistles.

Reverse.—A crown, and between laurel branches :—

> Fuentes D'Onor, 5th May, 1811.
> Ciudad Rodrigo, 19th Jan., 1812.
> Badajoz, 6th April, 1812.
> Salamanca, 22nd July, 1812.
> Vittoria, 21st June, 1813.
> Pyrenees, 28th July, 1813.
> Nivelle, 10th Nov., 1813.
> Nive, 13th Dec., 1813.
> Orthes, 27th Feb., 1814.
> Toulouse, 10th April, 1814.
> ANDREW ALLAN.

Suspended from a bar, inscribed " Peninsula."

Riband : Crimson, blue edge.

RIFLE BRIGADE.

55.—Shooting badge :—Small Maltese cross, with lions rampant between the arms ; in the centre, a crown and garter, with " Rifle Brigade " inscribed.

Reverse.—A bugle in the centre, and " Marksman " on the garter.

Riband : Dark green, with black edge.

CASES VIII. & IX.

———

NAVY WAR MEDALS, 1793 TO 1840.

GOLD NAVAL MEDAL.

After Lord Howe's victory, 1st June, 1794, it was thought expedient that a Medal should be struck. They are two sizes; the larger one given to Flag Officers, Captains of the Fleet, and Commodores; the smaller one to Post-Captains, only one officer below that rank having received it.

Obverse.—A figure of Victory, standing on the prow of an antique galley, placing a wreath of laurel on Britannia, having at her side a round shield, with the crosses of the Union Banner; her right foot resting on a helmet; a spear in her left hand.

Reverse.—A wreath, composed of oak and laurel, within which was the recipient's name and rank, and the event for which the medal was granted, and date.

It was worn—by the Admirals who received it for the 1st June, 1794—from a gold chain round the neck, but subsequently with a white riband, with blue edge.

John Inglis, Esquire, Captain, His Majesty's ship the "Belliqueux," on the 11th October, MDCCXCVII. The Dutch Fleet defeated.

Her Majesty was graciously pleased to grant a Medal to the survivors of the Navy who had taken part in the different engagements from 1793 to 1840, by an order dated 1st June, 1847.

The clasps for this medal number over two hundred, and bear either the name (if a great one) of the action, or the name of the ship capturing or defeating a ship of the enemy; and there were also clasps for boat-actions, with " boat-service " and date on clasp.

Obverse.—Head of Her Majesty. " Victoria Regina."

Reverse.—Britannia seated on a sea-horse, holding a trident in her right hand ; in her left, an olive branch.

Riband : White, with blue border.

One Clasp.

1.— 1st June, 1794 … … … THOMAS ROBSON.
CASE XII., No. 43.

2.— 1st June, 1794 … … … ALEXANDER BOYLE.
Lord Howe defeated the French Fleet, and captured six and sunk one sail of the line.

3.— " BLANCHE," 4th January, 1795 … HENRY GREELEY.
Captain Faulkner captured the French frigate " Pique."

4.— 14th March, 1795… … … JOHN WARD.
Vice-Admiral Hotham captured two sail of the French Fleet.

5.— 23rd June, 1795 … … … JOSEPH SKINNER.
Admiral Lord Bridport defeats the French, and captures three sail.

6.— " DRYAD," 13th June, 1796 … EDWARD VERLING.
Lord A. Beauclerk captures the French frigate " Proserpine."

7.— " AMAZON," 13th January, 1797 … JOHN BROWN.
Destruction of French 74, " Droits de l'Homme."

8.— ST. VINCENT, February, 1797 … THOMAS HALL.
Sir J. Jervis (Earl St. Vincent) defeats the Spanish Fleet, and captures four sail.

9.— " NYMPHE," 8th March, 1797 … JOHN COOK.
Capture of " Resistance " and " Constance."

10.— " CAMPERDOWN," 11th Oct., 1797 GEORGE CARTER.
Admiral Duncan defeats the Dutch Fleet.

11.—Isle St. Marcou, 6th May, 1798 John Campbell.
Defence of the Island.

12.—" Lion," 15th July, 1798   ... James Bowdry.
Captain Manby Dixon defeats four Spanish frigates, and captures
the " Santa Dorotea."

13.— Nile, 1st August, 1798 ...   ... John Smith.
Lord Nelson defeats the French Fleet in Aboukir Bay.

14.—12th October, 1798   ...   ... Jacob Michael.
Sir J. B. Warren defeats the French Squadron, captures the
" Hoche," 74, and two frigates.

15.—" Fisgard," 20th October, 1798 ... George Bright.
Captain Martin captures the French frigate " L'Immortalité."

16.—" Sybille," 28th February, 1799 Joseph Wright, Volr.
Captain Cook captures the French frigate " La Forte."

17.—Schiermonnikoog, 12th Aug., 1799 John Feary.
Attack on, and capture of the brig " Crash," 12 guns.

18.—Seine, 29th August, 1800   ... James Fitzgerald.
Captain Milne captures the French frigate " Vengeance."

19.—" Phœbe," 19th February, 1801 W. Ward.
Captain Sir R. Barlow captures the French frigate " Africaine."

20.—*Egypt, 8th March, 2nd Sept., 1801 William Parker, Lieut.
CASE XII., No. 37.

---

* Lieutenant Parker received the Turkish Gold Medal for his services during
this campaign (*vide* " O'Byrne's Naval Biographical Dictionary." He had also
the Medal with three clasps for " 14th March, 1795," " St. Vincent," and " Nile."
He died at Haslar Hospital, 24th July, 1862, aged 82.

21.— EGYPT, 8th March, 2nd Sept., 1801   FRANCIS SADGROVE.
Services on the Coast of Egypt.

22.— COPENHAGEN, 1801   ...   ...   SAMUEL MULLARD.
Lord Nelson defeats the Danes.

23.— " SPEEDY," 6th May, 1801   ...   DAVID GRAY.
Lord Cochrane.   Capture of " Gamo."

24.— GUT OF GIBRALTAR, 12th July, 1801   WILLIAM ROCHE.
Sir J. Saumarez.   Action with the French and Spanish Fleets ;
destruction of two Spanish ships, 112 guns each ; capture of
" St. Antonio," 74 guns.

25.— " CENTURION," 18th Sept., 1804··   JOHN SYMES, Quarter-
                                               master's Mate.

Action with " Marengo," Atalante " and " Semillante."

26.— " ACHERON," 3rd February, 1805   JOHN SIMPSON, Mdshpmn.
Gallant protection of 28 British merchant vessels, when attacked
by two 40-gun frigates.

27.— TRAFALGAR, 21st October, 1805   HENRY HITCHENS.
Lord Nelson's last victory.

28.— 4th November, 1805   ...   ...   JOHN M. DENNEY.
Sir R. Strachan captures four French sail of the line.

29.— ST. DOMINGO, February, 1806 ...   PATRICK BURNS.
Vice-Admiral Sir J. Duckworth captures four French sail of the
line.

30.— " AMAZON," 13th March, 1806 ···   HENRY COMFORT.

31.— " LONDON," 13th March, 1806 ···   J. T. CARDEN, Lt. R.N.
Sir H. Neale and Lieutenant Parker capture the " Marengo" and
" Belle Poule.'

32.—" BLANCHE," 19th July, 1806 ... THOMAS WEBBER.
Captain Lavis captures the " Guerrière."

33.—" ARETHUSA," 23rd August, 1806 SIMON FRAZER.
Captain Brisbane captures the Spanish frigate " Pomone."

34.—" HYDRA," 6th August, 1807 ... JOHN BENNETT.
Captain Mundy. At Begier, attack on batteries, and capture of
" L'Eugène " and " Caroline."

35.—" SAPPHO," 2nd March, 1808 ... WILLIAM HOWES.
Commander J. Langford captures the Danish brig " Admiral
Yawl."

36.—SAN FIORENZO, 8th March, 1808 ... JOHN PACEY.
Captain Hardinge captures the French frigate " Piedmontaise."

37.—" STATELY," 22nd March, 1808 ... WEBB HALL.

38.—" NASSAU," 22nd March, 1808 ... WM. ELLIS.
George Parker and R. Campbell destroy the Danish line-of-battle
ship " Prince Christian Frederick."

39.—" RAPID," 24th April, 1808 ... HENRY BAUGH.
Lieut. Baugh destroys Spanish gun boats at Faro.

40.—" VIRGINIE," 19th May, 1808 ... THOS. HOLBROOK.
Captain Brace captures the Dutch frigate " Guelderland.

41.—" SEAHORSE " with " BADERE ZAF-
        FER," 6th July, 1808 ... WILLIAM THOMSON.

John Stewart captures the Turkish frigate "Badere Zaffer."

42.—" CENTAUR," 26th August, 1808 .. JAMES GOSLIN.

Rear-Admiral Sir S. Hood defeats the Russian Fleet, and captures
the " Sewolod," 74 guns.

43.—" CRUIZER," 1st November, 1808... JAMES R. FORREST.

Lieutenant T. Wells defeats flotilla off Gottenburgh, and captures a brig.

44.—"AMETHYST" with "THETIS," 10th
        November, 1808 ... ... ISAAC SMITH.

Sir M. Seymour captures the French frigate " Thetis."

45.—MARTINIQUE, February, 1809 ... SIR C. F. SMITH.
                                        CASE XII., No. 34.

46.—MARTINIQUE, February, 1809 ... WILLIAM NEALE.

Admiral Sir A. Cochrane captures Martinique.

47.—" HORATIO," 10th February, 1809 JOSEPH ALLEN.

Commander George Scott. Capture of the French frigate " Junon."

48.— BASQUE ROADS, 1809 ... ... S. SPARSHOTT, Master's
                                        Mate.

Destruction of ships in Basque Roads.

49.— " CASTOR," 17th June, 1809 ... W. ANDREWS.

Commander Roberts. Capture of the French ship " Hautpoolt."

50.—GUADALOUPE, Jan. and Feb., 1810 MICHAEL FLYNN.

Vice-Admiral Sir A. Cochrane captures the island from the French.

51.— " SPARTAN," 3rd May, 1810 ... JOSEPH ROGINSON.

Commander Japhet Brenton defeats French frigates, and captures the " Sparvière."

52.— AMANTHEA, 25th July, 1810

Capture of transports off Amanthea.

53.— BANDA NEIRA, 9th August, 1810... WILLIAM FRANCIS.

Capture of the island from the Dutch.

54.—" Boadicea," 18th Sept., 1810 ...   H. S. Clifford, Lt. R.N.
Action with the French frigate " Venus," and re-capture of the
" Ceylon."

55.—Lissa, 13th March, 1811   ...   James Penton.
Action and capture of Franco-Venetian frigates off Lissa.

56.—Anholt, 27th March, 1811   ...   William Wettersby.
Captain J. W. Maurice.   Defence of the Island of Anholt.

57.— Off Tamatave, 20th May, 1811 ...   John Calder.
Action with French frigates, and capture of the " Renommée "
and " Neréide."

58.— Java, Aug. and Sept., 1811   ...   James Noakes.
Rear-Admiral Honourable R. Stopford.   Capture of the island
from the Dutch.

59.—" Locust," 11th Nov., 1811   ...   Samuel Bachel.
Action with flotilla and capture of a gun brig off Boulogne.

60.—" Victorious" with " Rivoli" 22nd
        Feb., 1812 ...   .   C. Arnold.

61.—" Weazel," 22nd Feb., 1812 ...   William Young.
Captain Talbot and W. Andrew capture the French ship " Rivoli,"
74 guns.

62.—" Northumberland," 22nd May,
        1812 ...   .   Robert Dillon.
Destruction of two French frigates and a brig.

63.—Off Mardoe, 6th July, 1812   ...   James Tornton.
Destruction of two Danish frigates and two brigs.

64.— ' Shannon" with "Chesapeake,"
     1st July, 1813 ... ... *Joseph Johnston.
Captain Brooke captures the American frigate "Chesapeake."

65.— "Pelican," 14th Aug., 1813 ... William Baker.
Commander Maple captures the American brig "Argus."

66.— St. Sebastian, Aug. and Sept., 1813 H. Richmond, Mdshpmn.
In support of the Army.

67.— "Venerable," 16th Jan., 1814... Thomas Wilson.
Rear-Admiral Durham captures two French frigates.

68.— "Eurotas," 25th Feb., 1814 ... John Mees.
J. Phillimore captures the French frigate "Clorinde."

69.— "Hebrus" with "L'Etoile,"
     27th March, 1814 ... ... Thomas Jones.
E. Palmer captures the French frigate "L'Etoile."

70.— "Phœbe," 28th March, 1814 ... Thomas Lark.
James Hillyar captures the American frigates "Essex" and "Essex Junior."

71.—The Potomac, 17th Aug., 1814 John Powell.
At Alexandria in America, and destruction of shipping in the Potomac.

72.— "Endymion" with "President,"
     15th Jan., 1815 ... ... James Richardson.
H. Hope captures the American frigate "President."

73.— Gaeta, 24th July, 1815... ... John Wilson.
Attack on and reduction of Gaeta.

---

  * This man formed one of the boarding party on this occasion, and was wounded.

74.— ALGIERS, 27th August, 1816 ... THOMAS HAUSLER.
Admiral Lord Exmouth bombards Algiers.

75.— NAVARINO, 20th October, 1827 ... DAVID HAWKINS.
Vice-Admiral Sir Edward Codrington defeats the Turkish Fleet.

76.— SYRIA, November, 1840 ... ... MAJOR-GENERAL SIR C.
F. SMITH, R.E.
CASE XII., No. 84.

77.— SYRIA, November, 1840 ... ... HENRY BAILEY.

78.— SYRIA, November, 1840 ... ... T. LEACH.
CASE XII., No. 89.

Admiral the Honorable Sir Robert Stopford captures Acre. And
for operations on the coast of Syria.

Two Clasps.

79.—1st June, 1794 and 23rd June, 1795 EDWARD BURT.

80.—BOAT SERVICE, 17th March, 1794,
"AMAZON," 13th March, 1806... JAMES SMITH.
1.—Boats of Fleet under Sir John Jervis, board and capture
the French frigate, "Bleuvenne," and others, in Fort
Royal Bay.
2.—Capture of the "Marengo" and "Belle Poule."

81.—ST. DOMINGO, 1st June, 1794 ... RICHARD GOFF.

82.—St. Vincent, 14th March, 1795 ...   John Cameron.

83.—Lowestoft, 24th June, 1795, and
    Egypt ...   .   Gilbert White.
  1.—Action with the French frigates, " Minerve " and
    " Artemise." Capture of the former.

84.—" Indefatigable," 20th April,
    1796, and " Indefatigable,"
    13th January, 1797 ...   ...   Thomas Groube.
Captain Sir E. Pellew captures the French frigate " Virginie,"
and assists at the destruction of the French ship " Droits de
l'Homme," 74.

85.—Camperdown, Egypt   ...   ...   William Clark.

86.—Camperdown, Copenhagen   ...   Thomas Griffiths.

87.—Camperdown, Lissa   ...   ...   Edward Edwards.

88.—" Mars," 21st April, 1798 ;
    " Comus," 15th August, 1807 ...   Thomas Saunders.
  1.—Alexander Hood captures the French 74 " L'Hercule."
  2.—Captain E. Heywood captures the Danish frigate
    " Frederickscoarn."

89.—Nile, Egypt   ...   .   Francis Wade.

90.—Nile, " Shannon " with " Chesa-
    peake " ...   .   Michael Pealeng.

91.—Egypt, 12th October, 1798   ...   Richard Bebb.

92.—Acre, 80th May, 1799 ; Egypt ...   Charles Taylor.
  1.—For defence of Acre during the siege.

98.—Egypt, Trafalgar   ...   ...   Samuel Green.

94.—EGYPT, BASQUE ROADS, 1809 ... J. ANDERSON, Carpenter.

95.—EGYPT, Boat Service, 1st November, 1809 ... ... JAMES MURPHY.

96.—EGYPT, ALGIERS ... ... THOMAS BISHOP.

97.—COPENHAGEN, 1801; JAVA ... SAMUEL WEBB.

98.—TRAFALGAR, ST. DOMINGO ... J.B. COUSINS, Mdshpmn.

99.—TRAFALGAR, " SIRIUS," 17th April, 1806 ... ... JOHN HENNESSEY.

   2.—Captain Prowse. Action with French flotilla at Civita Vecchia; capture of the " Bergère."

100.—TRAFALGAR, Boat Service, 16th July, 1806 ... ... WM. McGINN.

101.—TRAFALGAR, BASQUE ROADS, 1809 WILLIAM CHINNARY.

102.—TRAFALGAR, Boat Service, 25th July, 1809 ... ... EDWARD BURN.

   2.—Capture of Russian gunboats in Aspo Roads.

103.—TRAFALGAR; " HEBRUS " with " L'ETOILE " ... ... WILLIAM SMITH.

104.—TRAFALGAR, Boat Service, 6th May, 1814 ... ... WILLIAM WILLS.

   2.—Action at Oswego, on Canadian Lakes. We lost one Captain of Marines and 17 men killed; 6 officers, 57 men wounded.

There is no authority for this clasp.

105.—SAN FIORENZO, 14th Feb., 1805; SAN FIORENZO, 8th Mar., 1808 GEORGE BARNEY.

   1.—Captain Lambert captures the French frigate, " Psyche."
   2.—Captain Hardinge captures the French frigate " Piedmontaise."

106.—Basque Roads, 1809; 4th Nov.,
1805. ...   .   Richard Boley.

107.—Navarino, 4th Nov., 1805   ...   Samuel Lawrence.

108.—Trafalgar, Martinique ...   ...   John Chambers.

109.—Martinique, 4th Nov., 1805   ...   William Harris.

110.—St. Domingo; Off Mardoe, 6th
July, 1812 ...   .   James   P.   Stewart,
Capt , R.N

Captain Stewart, of the "Dictator," commanded the ships off Mardoe.

111.—Boat Service, 16th July, 1806;
"Centaur," 26th Aug., 1808   Daniel Clarke.

112.—"Arethusa," 23rd Aug., 1806;
Curacoa   ...   .   James Smith.

2.—Capture of the Island of Curacoa, 1st January, 1807.

113.—"Amethyst" with "Thetis;"
"Amethyst," 5th April, 1809   Christian Lauterburgh.

2.—Sir M. Seymour captures the "Niemen."

114.—Off the Pearl Rock, 13th Dec.,
1808; Martinique   ...   ...   George Harding.

1.—Action with batteries and capture of a corvette.

115.—St. Domingo, Martinique   ...   Thomas Thomson.

116.—Martinique, Guadaloupe   ...   Wm. Bone.

117.—Basque Roads, 1809;
"Thunderer," 9th Oct., 1813   David Finn.

2.—Captain Pell captures the "Neptune," an armed lugger.

118.— Anse le Barque, 18th Dec., 1809 ;
      Guadaloupe   ...   ...   James Bell.

  1.—Storming batteries at Anse le Barque, and capture of the "Loire" and "Seine" frigates.

119.— Guadaloupe, "Shannon" with
      "Chesapeake"   ...   ...   John F. Winnister.

120.— Guadaloupe, Algiers ...   ...   Thomas Dillon.

121.— Banda Neira, Java   ...   ...   Robert Page.

122.— Lissa, Pelagosa, 29th Nov., 1811 Thomas Johnson.

  2.—Capture of two French frigates.

123.— Basque Roads, 1809 ; Java   ...   John Taylor.

124.— Lissa, Algiers ...   .   William Mitford.

125.— Off Tamatare, 20th May, 1811 ;
      "Phœbe," 28th March, 1844   William Morgan.

126.— Pelargosa, 29th Nov., 1811 ;
      Algiers   ...   .   John Williams.

127.— "Gaeta," 24th July, 1815,
      Algiers   ...   .   James Berry.

128.— Navarino, Syria   ...   ...   James Coker.

129.— Algiers, Navarino   ...   ...   John Rowe.

Three Clasps.

180.—1st June, 1794, St. Vincent, Nile   James Kite.

181.—Trafalgar ; 1st June, 1794 ;
        Amazon, 13th March, 1806 ...   James Nipper.

182.—14th March, 1795 ; Lowestoft,
        24th June, 1795 ; "Scorpion,"
        31st March, 1804   ...   ...   George Salvedore.

  3.—Captain Hardinge attacks and captures vessels in the Vlie
        Roads.

183.—14th March, 1795, St. Vincent ;
        Nile   ...   .   Wm. Parker, Midship-
                                              man.
                                              Case XII., No. 37.

184.—23rd June, 1795, St. Vincent ;
        Egypt   ...   .   John Ormiston.

185.—St. Vincent, Nile, Copenhagen,
        1801   ...   .   Joseph Stevenson.

186.—St. Vincent, St. Domingo, Guada-
        loupe   ...   .   Francis Budd.

187.—"Harpy," 5th Feb., 1800 ; Copen-
        hagen, 1801 ; Guadaloupe ...   William Talbot.

  1.—Action with the French frigate "Pallas," and her conse-
        quent capture by the "Loire."

188.— Egypt Curacoa, Algiers   ...   Jonathan Rand.

139.— Copenhagen, 1801 ; Martinique,
     Algiers   …   …   … Josh. Groscott.

140.— Trafalgar, St. Domingo, Java Abraham Coop.

141.— St. Domingo, Martinique, Java Samuel Jeffry.

142.— Martinique ; Pompee, 17th June,
     1809 ; Guadaloupe ···   … Joseph Morgan.
  2.—Captain Fahie. Chase and capture of the French frigate
    " Hautpoolt."

143.— Basque Roads, 1809 ; Boat Ser-
     vice, 23rd Nov., 1810 ; Algiers John Smith.

144.— Lissa ; Pelagosa, 29th Nov.,
     1811 ; Syria ···   …   … William Tuckey.

145.— Algiers, Navarino, Syria   … John Bradley.

Four Clasps.

146.— 12th October, 1798, St. Domingo,
     Java, Algiers..   …   … John Boon.

147.— Copenhagen, 1801 ; 4th Nov.,
     1805 ; Basque Roads, 1809 ;
     Boat Service, 27th Sept., 1810 John Mann.
  4.—Boats of " Caledonia," " Valiant," and " Armide,"
    storm batteries at Pointe de Ché, Basque Roads.

148.— Copenhagen, 1801 ; Basque
     Roads, 1809 ; The Potomac,
     17th August, 1814 ; Algiers John Ward.

Six Clasps.

149.— 14th March, 1795 ; " Minerve," 19th Dec., 1796 ; St. Vincent ; Egypt ; Martinique ; Boat Service, 29th April, 1813.

George Cockburn, Rear-Admiral.

Captain Cockburn commanded the " Meleager " on 14th March, 1795.

" Minerve," 19th December, 1796. Commodore Nelson. Captain George Cockburn.

He commanded the " Minerve " at St. Vincent, and on the Coast of Egypt.

Commodore George Cockburn commanded the " Pompée " at Martinique ;

And, as Rear-Admiral Sir George Cockburn, personally commanded the boats of the Fleet, 29th April, 1813, at French Town and Havre de Grâce.

BOAT ACTIONS.

One Clasp.

150.— Boat Service, 17th March, 1794 JAMES LUNN.

The boats of the ships under Sir John Jervis capture the " Bienvenue," and other vessels in Fort Royal Bay.

151.— Boat Service, 9th June, 1799... GEORGE HARDING.

Boats of the " Success," under Lieutenant P. Facey, capture a Spanish polacca.

152.— Boat Service, 29th August, 1800 THOMAS LAMBERT.

Boats, under Lieutenant Burke, cut out the " Guêpe," of 18 guns.

153.— Boat Service, 16th July, 1806.. CHARLES FOOT.

Boats, under Lieutenant Sibley, cut out the " Cæsar."

154.— Boat Service, 21st January, 1807 WILLIAM MILLS.

Boats of the " Galatea," under Lieutenant Coombe, capture the " Lynx."

155.— Boat Service, 7th July, 1809 ... GEORGE KING.

Boats, under Lieutenant J. Hankey, destroy convoy and gunboats at Hango Head, Baltic.

156.— Boat Service, 1st Nov., 1809 ... FRANCIS DOVE.

Boats, under Lieut. Tailour, capture eleven armed vessels in the Bay of Rosas.

157.— Boat Service, 13th Feb., 1810 ... JOHN JONES.

Boats, under Lieut. Guion, attack nine and capture one French gunboat.

158.— Boat Service, 1st May, 1810 ... THOMAS GRIFFIN.

Boats of the " Nereide," under Commander Willoughby, capture the Fort at Jacotel.

159.— Boat Service, 23rd Nov., 1810 ... GEORGE HILL.

Bomb and mortar Fleet, and boats of the Cadiz Fleet, under Captain Sir D. Hall, attack and destroy the shipping at Port St. Mary.

160.— Boat Service, 2nd Aug., 1811 ... THOMAS HARE.

Boats, under Lieut. S. Blyth, capture three Danish gun-brigs in the River Jahde.

161.— Boat Service, April & May, 1813 DAVID GREENBERRY.

Boats of the Fleet, under the command of Rear-Admiral Sir George Cockburn, attack French town and Havre-de-Grâce, and destroy the Fort and Iron Foundry.

162.— Boat Service, 2nd May, 1814 ... JAMES CLARE.

Boats, under Lieut. Shaw, at Morgion, blow up battery and destroy six vessels.

163.— Boat Service, 24th May, 1814... WILLIAM AARON.

Boats of the " Elizabeth," under Lieut. M. Roberts, capture the " L'Aigle."

164.— Boat Service, 14th Dec., 1814 ... JAMES BRILL.

Boats, under Captain Nicholas Lockyer, capture five gun-vessels and a sloop.

Two Clasps.

165.— $\left\{\begin{array}{l}\text{Boat Service, 16th July, 1806}\\\text{Boat Service, 24th May, 1814}\end{array}\right\}$ JOHN DEARNENS.

CASE X.

NAVAL MEDALS,

SINCE 1840.

The Naval War Medals for services since 1840, are in most instances the same as those issued to the Army, and are worn with the same riband.

INDIA MEDAL, 1799-1826.

CLASP, " AVA."

1.—R. Bell, A.B.

1st CHINA, 1842.

2.—William Reeves ... ... H.M.S. "Blonde."

CAMPAIGN IN SCINDE, 1843.

MEANEE.

3.—T. Trimlett, A.B.... ... ... H.C.V. "Planet."

HYDERABAD.

4.—John Seaton, O.S.... ... ... E.I.C.S. "Comet."

CAMPAIGN IN THE PUNJAB, 1848-1849.

NO CLASP.

5.—T. Rivett, 2nd Class Engineer... Ind. Flot.

MOOLTAN.

6.—Richard Burns, A.B. ...   ... Ind. Flot.

There were only fourteen with this Clasp issued to the Navy.

SOUTH AFRICA, 1834-1835.

7.—Joseph Ward.

8.—J. Curtis, A.B.

Case XII., No 38.

CRIMEA, 1854-1855.

NO CLASP.

9.—James Coleman, Boy, 1st Class.

One Clasp.

SEBASTOPOL.

10.—*John Starling, Boatswain ... H.M.S. " Sampson."

11.—W. Andrew Nisbiat ...   ... 75th Co., R. M. L. I.

---

* This Medal was given to J. Starling by Her Majesty on the Horse Guards Parade, 18th May, 1855.

AZOFF.

12.—J. Jeffrys, Boy, 1st Class.

This Clasp was only given to the Navy, and not to the Army.

Two Clasps.

BALACLAVA, SEBASTOPOL.

13.—S. Viney   ...   . 20th R. M., H.M.S.
"Albion."
Case XII., No. 40.

INKERMAN, SEBASTOPOL.

14.—George Yule, Sergeant   ... R. M. A.
Case XII., No. 42.

15.—Joseph Ward ...   . H.M.S. "Bellerophon."
Case XII., No. 38.

16.—J. Walker, Corporal ...   ... 11th Co., R. M. A.

AZOFF, SEBASTOPOL.

17.—Michael J. Hanlon, A.B.   ... H.M.S. "Albion."

Three Clasps.

INKERMAN, BALACLAVA, SEBASTOPOL.

18.—Edgar Baughan, Private   ...   Royal Marines.

BALTIC, 1854-1855.

Obverse.—Head of Her Majesty. " Victoria Regina."
Reverse.—Britannia seated, holding in her right hand a trident, looking towards the distant fortress of Bomersund. Above, the word " Baltic, 1854, 1855."
Riband : Yellow, light blue border.

19.—James Bute, Sergeant   ...   Royal Marines.
Case XII., No. 41.

20.—G. W. Beachey   ...   ...   H.M.S. " Hawke."

INDIA GENERAL SERVICE MEDAL.

PERSIA.

21.—R. Dobbin, A.B.   ...   ...   " Punjaub," S.F.

PEGU.

22.—W. Mitchell, O.S.   ...   ...   " Falkland " Sloop.

INDIA MUTINY, 1857-1858.

NO CLASP.

23.—B. Macdonald, Seaman   ...   In Brigade.

One Clasp.

LUCKNOW.

24.—E. Meacher, A.B.   ...   ...   "Shannon."

Two Clasps.

LUCKNOW; RELIEF OF LUCKNOW.

25.—Thos. Honeybourne, Ord.   ...   "Shannon."

2nd CHINA, 1857—1860.

NO CLASP.

26.—E. J. Bryant   ...   .   H.M.S. "Scout."

One Clasp.

FATSHAN, 1857.

27.*—John O'Brien ...   .

TAKU FORTS, 1858.

28.—James Henry, A.B.   ...   ... H.M.S. " Pique."

CANTON, 1857.

29.—J. Liddon, Drummer, P.D.   Royal Marines.

Three Clasps.

CANTON, 1857; TAKU FORTS, 1860; PEKIN, 1860.

30.—John Beck   ...   . Royal Marine Light Inf.

* This clasp was only given to the Navy.

𝔉𝔦𝔳𝔢 ℭ𝔩𝔞𝔰𝔭𝔰.

FATSHAN, 1857; CANTON, 1857; TAKU FORTS, 1858; TAKU FORTS, 1860; PEKIN, 1860.

31.—Thomas Cole ...   .   Royal Marine Artillery.
This is the only Medal for China with five clasps that was ever issued.

Extract from a letter from Colonel Williams, R.M.A. :—

"1st Co. Gunner Thomas Cole was invalided 18th May, 1865, and was then in possession of the China Medal, with five clasps, having been present at Fatshan, Capture of Canton, Taku Forts, 1858–60, and also at the operations against Pekin.

(Signed)   " J. W. C. Williams,

" Colonel-Commandant."

NEW ZEALAND.

1845—1846.

32.—J. Bailey, A.B.   ...   ...   H.M.S. "Castor."

1846—1847.

33.—T. Coombes, A.B.   ...   ...   H.M.S. "Calliope."

1847.

34.—J. Turner, Ropemaker   ...   H.M.S. "Inflexible."

1860—1861.

35.—J. Oakes, Gunner, R.M.A.   ...   H.M.S. "Pelorus."

178

1863—1864.
36.—J. Martin. Boy, 1st Class   ...   H.M.S. " Esk."

1863—1865.
37.—Robert Course, Gunner, R.M.A.   H.M.S. " Eclipse."

ABYSSINIA.

1868.
38.—H. Triggs, Musician   ...   ...   H.M.S. " Argus."

ASHANTEE.

1873—1874.
39.—J. Walters, A.B.   ...   ...   H.M.S. " Himalaya,'
1873–1874.

𝔒𝔫𝔢 𝔆𝔩𝔞𝔰𝔭.

COOMASSIE.

1873—1874.
40.—T. G. Hart, Ordinary ...   ...   H.M.S.   " Active,"
1873–1874.

LONG SERVICE AND GOOD CONDUCT (1).

This Medal was instituted by King William IV., August 24th, 1831, for seamen and marines.

Obverse.—A crown and anchor, encircled by a wreath of oak.
Reverse.—The name, rating, ship, and length of service of the recipient, within a circle containing the words, " For long service and good conduct."
Riband : Dark blue.

41.—Wm. Button, Captain of foretop   H.M.S. " Pilot," 21 yrs.
42.—W. Patchet, Private   ...   ...   Royal Marines, H.M.S. " Blonde," 22 years.

LONG SERVICE AND GOOD CONDUCT (2).

This Medal was substituted by Her Majesty for the last.

Obverse.—Head of Her Majesty. " Victoria Regina."
Reverse.—A line-of-battle ship at anchor, surrounded by a cable forming a circle, within which are the words, " For long service and good conduct."

The recipient's name is inscribed on the edge.

Riband : Blue, white edge.

43.—J. A. Bute, Sergeant ...   ...   19th Co. R.M., 32 yrs.
       Case XII., No. 41.

44.—C. Viney, Private   ...   ...   20th Co. R.M., H.M.S. " Albion," 20 years.
       Case XII., No. 40.

45.—T. Leach, Master-at-Arms   ...   H.M.S. " Victoria and Albert," 22 years.
       Case XII., No. 39.

46.—James Hunt, Boatswain's Mate   H.M.S. " Trafalgar," 23 years.

47.—Henry Salter, Corporal   ...   R.M., H.M.S. " Merlin," 22 years.

ENGINEERS' GOOD CONDUCT MEDAL.

This Medal was issued by authority of the Admiralty, in 1842: "The Lords Commissioners of the Admiralty have caused a Medal to be struck, to be presented as rewards to engineers of the first class, serving in Her Majesty's Navy, who, by their good conduct and ability, deserve some special mark of notice." It was discontinued in 1847, when the relative rank of engineers was raised.

There were only seven ever issued.

Obverse.—A two-masted paddle-wheel steamer.

Reverse.—"For ability and good conduct." A crown and anchor within a circle, containing the rank and name of recipient.

Riband : Blue, white edge.

48.—Mr. Samuel Bult Meredith ... 2nd Class Engineer.

MERITORIOUS SERVICE.

Same Medal as the Army Medal given to sergeants of Marines; worn by them with a blue riband instead of red.

49.—James Bute, Sergeant... ... R.M., 8th Dec., 1852.
Case XII., No. 41.

50.—(No name.)

CONSPICUOUS GALLANTRY

This Medal was instituted by an Order in Council, dated 13th August, 1855, and bestowed on petty officers, seamen, and sergeants, corporals, and privates of Royal Marines who particularly distinguished themselves in action with the enemy, in the proportion of eight petty officers, or sergeants or corporals of Marines, and ten seamen or privates of Marines for every thousand men engaged, with gratuities of £15, £10, and £5, the amount not to exceed £4,000 in any one year.

51.—John Taylor, Captain of the
    Forecastle   ...   .   H.M.S. London.

He received this medal for conspicuous gallantry at Inkerman.

VICTORIA CROSS.

Same as the Army, but worn with a blue riband.

52.—E. St. John D—— ... ... Midshipman.
    5th Nov., 1854 ; 18th June, 1855.

London Gazette, 24th Feb., 1857.

Sir Stephen Lushington recommends the officer :

1st.—For answering a call for volunteers to bring in powder to the battery, from a waggon in a very exposed position under a destructive fire, a shot having disabled the horses. (This was reported by Captain Peel, commanding the battery.)

2nd.—For accompanying Captain Peel at the battle of Inkerman as Aide de Camp.

3d.—For devotion to his leader Captain Peel, on the 18th June, 1855, in tying a tourniquet on his arm on the glacis of the Redan, whilst exposed to a very heavy fire. (Dispatch from Sir S. Lushington enclosed in letter from Admiral Lord Lyons, 10th May, 1856.)

DAVISON'S " NILE."

This Medal was bestowed by a private gentleman after the victory of the Nile : Mr. Davison, Lord Nelson's agent.

Obverse.—A figure standing upon a rock, holding an olive branch and displaying the bust of Nelson on a shield. Around the bust, " Europe's Hope and Britain's Glory." Behind the figure, an anchor. " Rear Admiral Lord Nelson of the Nile."

Reverse.—The British Fleet sailing into Aboukir Bay. " Almighty God has blessed His Majesty's Arms." " Victory of the Nile, August 1st, 1798." On the edge, " From Alex. Davison, Esq., St. James's Square. A tribute of regard."

53.—Copper gilt.

54.—Bronze.

BOULTON'S " TRAFALGAR."

Mr. Boulton, the venerable proprietor of Soho, has solicited the permission of Government that he might be allowed to strike a medal at his own expense, in commemoration of the brilliant victory off Cape Trafalgar, and to present one to every seaman who served that day on board the British Fleet. The permission was immediately granted.—*Naval Chronicle.*

Obverse.—Bust of Nelson. " Horatio Viscount Nelson, K.B., Duke of Bronte."

Reverse.—Battle of Trafalgar. " England expects every man to do his duty." " Trafalgar, October 21st, 1805."

Riband : Blue.

55.—Silver.

LORD ST. VINCENT'S TESTIMONY OF APPROBATION.

This Medal was given by Earl St. Vincent.

Obverse.—Earl St. Vincent, surrounded by a laurel wreath. " Earl St. Vincent's testimony of approbation, 1800."

Reverse.—A sailor and marine joining hands, within a laurel wreath. The outlines of the crosses of the Union Banner. " Loyal and True," with crown and laurel wreath.

56.—Silver Gilt.

57.—Silver.

TAYLEUR FUND MEDAL.

58.—*Obverse.*—A ship in a storm. " Tayleur Fund for the succour of shipwrecked strangers."

Reverse.—" To John Newnham, H.M.S. ' Ajax.' For distinguished gallantry in saving life at Kingstown, 9th February, 1861." " The Right Honorable Lord Talbot of Malahide, Chairman."

Riband : Blue, white edge.

On 24th January, 1854, the emigrant ship " Tayleur" was totally lost off Howth, with 380 lives. Such a large subscription was raised that a surplus was left, which was converted into a fund for granting rewards and medals for efforts made to save life from shipwreck.

ROYAL NATIONAL INSTITUTION FOR THE PRESERVATION OF LIFE FROM SHIP-WRECK.

59.—*Obverse.*—Head of George IV., 1824. " Royal National Institution for the Preservation of Life from Shipwreck." " George the Fourth, patron."

Reverse.—Three men in a boat in the act of saving a man. " Let not the deep swallow me up."

Riband : Blue.

Lieutenant NATH. NEWNHAM, R.N., voted December 17th, 1840.

ARCTIC DISCOVERY, 1818–1855.

The Warrant for this Medal is dated January 30th, 1855; to all engaged in the several expeditions to discover a North-West

Passage under Ross, Parry, Franklin, Collinson, and McLure, between 1818–55.

An octagonal medal.

Obverse.—Head of the Queen. " Victoria Regina."

Reverse.—A ship between icebergs; in the foreground, sailors and sledge; " 1818, 1855." Suspended from a star.

Riband: White.

 60.—James Bute, Sergeant ... Royal Marines.
Case XII.

 61.—Joseph Busbridge.

2ND ARCTIC DISCOVERY, 1876.

This Medal was granted by Her Majesty to the expedition under Sir George Nares, in the " Discovery " and " Alert."

Obverse.—Diademed head of Her Majesty. " Victoria Regina, 1876."

Reverse.—A ship among floating blocks of ice.

Riband: White.

 62.—Jas. Hand, A.B. ... ... H.M.S. " Discovery.'

CASE XI.

MEDALS BESTOWED BY FOREIGN POWERS ON BRITISH OFFICERS AND MEN.

AFFGHANISTAN.

ORDER OF THE DOORANEE EMPIRE.

1.—A gold Maltese cross, on a larger one of silver, supported two crossed swords. In the centre, surrounded by a circle of pearls, a blue and green enamelled ground, in Persian capitals, is " Duri Dauran " (Pearl of the Age).

Riband : Crimson and green.

This Order belonged to Lord Keane.

It was instituted by Shah Soojah-ool-Moolk, on his restoration to his kingdom, to reward the British officers by whom it was effected.

CHINA.

GOLD MEDAL, RUBY BUTTON.

2.—*Obverse.*—Imperial arms.

Reverse.—In Chinese characters, " The Imperially-bestowed precious star of the first rank, Ta Tsing, the great Tsing (Manchow Dynasty, reigning line), Ya tsze paon sing fit tang."

Worn, suspended by a gold pin, on the breast. From a loop on the lower side of the medal, two green tassels are suspended, knotted, and tied in pink and red.

GOLD MEDAL, BLUE BUTTON.

3.—Same as above.

CRYSTAL BUTTON.

4.—Silver.

Obverse.—Scrolls.

Reverse.—In Chinese characters, "For encouragement of merit . Military valour. Pai tsing kung mo."

Riband : Yellow silk braid, with shaded blue threads round it at the extremity, by which, from a loop, the medal is worn. From a similar loop, attached to the lower part of the medal, are suspended two green tassels, knotted with pink, This medal belonged to Quartermaster-Sergeant H. Marfleet, C Battery, 14th Brigade, Royal Artillery. Her Majesty's permission to accept and wear this decoration was granted on December 14th, 1864.

FRANCE.

LEGION OF HONOUR.

Obverse.—A white enamelled cross of ten points, resting on a wreath of laurel, suspended from an Imperial crown ; in the centre the head of Napoleon I., within a blue enamelled circle, inscribed " Napoléon Emp. des Français."

Reverse.—In centre, within a blue enamelled circle, bearing the legend " Honneur et Patrie," is the French eagle, with semi-out-spread wings.

Riband : Red.

This decoration was given to selected officers and men of the Army and Navy in the Crimea.

5.—Those in my collection were given to Sergeant-Major , 4th Dragoon Guards, afterwards Cornet and Adjutant 9th Lancers.

6.—The other to Serjeant George Yule, Royal Marine Artillery.
Case XII., No. 42.

FRANCE.

WAR MEDAL.

7.—*Obverse.*—The French eagle with outspread wings, holding in its talons a silver-gilt medal, in the centre of which are the words "Louis Napoleon," within a blue circle.

Reverse.—Within a wreath of laurel are the words "Valeur et Discipline."

Riband : Orange watered, with green edges.

This medal was given to 226 selected non-commissioned officers and men of the army and navy for service in the Crimea.

The medal in this collection belonged to Private John Goldsmith, 90th Light Infantry.

The only two officers who received this medal were Field-Marshal H.R.H. the Duke of Cambridge, K.G., and General Sir William Codrington, G.C.B., who was presented by Marshal Pelissier with his own medal.

FRANCE.

GENEVA CROSS.

8.—*Obverse.*—Bronze cross; "1870—1871; Societé Française de Secours aux Blessés des Armées de Terre et de Mer."

Reverse.—Jean Wilkinson.

Riband : White, with crimson cross.

HANOVER.

THE GUELPHIC ORDER.

9.—Founded by King George IV., when Prince Regent of England, in the name of his father, George III., on the 12th August, 1815, his birthday. This Order was discontinued when Hanover ceased to form part of this country.

The star is silver, of eight points; a white horse on a red enamelled ground, surrounded by a laurel wreath, and the motto " Nec aspera terrent; " the whole resting on two crossed swords.

Riband : Light blue, watered.

PAPAL MEDAL.

10.—An inverted cross resting on the head of a serpent; round the cross, on a scroll, " Pro Petri Sede, Pio IX., P. M. A. X., XV."

Reverse of Scroll.—" Victoria quæ vincit mundum fides nostra."

Riband : Crimson, with two white stripes with narrow yellow stripe on each side.

This medal was given by Pope Pius IX. There is no recognised authority to wear it, and it was only given to Irish Volunteers under Major O'Reilly.

PORTUGAL.

THE TOWER AND SWORD.

11.—A star of five points, enamelled white; the points connected by a wreath suspended from a tower.

Obverse.—A sword resting on an oak wreath, with the legend, " Valor, Lealdade e Merito " (Valour, Devotion, Merit).

Reverse.—An open book ; on one page, the Portuguese arms ; on the other, " Carta Constitutional da Monarquia," and the legend, " Pelo Rei e pela Lei " (For the King and the Law).

Riband : Dark blue.

SARDINIA.

Obverse.—The arms of Savoy and the crown of Sardinia, surrounded by a wreath of palm and olive branches, and the inscription, " Al Valore Militar."

Reverse.—Laurel branches, surrounded with the words "Spedizione D'Oriente, 1855–1856; " in the centre, the recipient's name.

Riband : Dark blue.

12.—WILLIAM CAMPBELL, Corporal ...　79th Regiment.
CASE XII., No. 25.

13.—G. SYMONS, Sergeant ...　　...　Royal Artillery.
CASE XII. No. 31.

14.—

This medal was given to 400 selected officers and non-commissioned officers and men of the Army, and 50 sailors and Marines during the Crimean War.

SPAIN.

ORDER OF CHARLES III.

15.—This Order was established by King Charles III., in 1771. The statutes now in force date from the 12th January, 1804, as promulgated by Charles IV. The Order was abolished by Joseph Bonaparte in 1808, but was restored in 1814. " It is devoted to the pure conception of the Virgin " and destined to reward marked zeal and ability displayed by the nobility for the interests of the crown.

Obverse.—On the centre of the cross is the image of the Virgin Mary, enamelled in proper colours and standing on a silver crescent, within a blue circle.

Reverse.—The initials of the founder, encircled by a laurel wreath, with the words " Virtute et Merito ; " worn from a ring chased with laurel leaves.

Riband : White watered, with blue edge.

" To Charles Felix Smith, Esquire, Lieutenant-Colonel in the Army and Captain of the Corps of Royal Engineers, in consideration of the military services rendered by him in the Peninsula, but more particularly for his skill and highly distinguished merit in the defence of Tarifa in December, 1811. Royal license granted 21st September, 1814. Knighted 10th November, 1814."

Case XII., No. 34.

ORDER OF ST. FERDINAND.

16.—This Order was instituted by the Cortes in 1811, in consequence of the old Orders having become articles of traffic, and was accessible to all grades of military men.

A cross of four equal arms enamelled white, and suspended from a green enamelled laurel wreath.

Obverse.—The image of St. Ferdinand, and, within a blue circle, the words " Al Merito Militar."

Reverse.—Two gloves under a crown, on a light blue ground, within a dark blue circle, with the words " La Patria."

Riband : Red, with orange edge.

This cross belonged to Sir C. F. Smith, R.E., and was conferred on him for services at Tarifa.

Case XII., No. 34.

ORDER OF MARIA ISABELLA LOUISA.

17.—A silver cross, in the centre of which are the letters M.I.L. in monogram ; suspended from a crown and ring.

Riband : Light blue.

This decoration was founded by Ferdinand VII., and was conferred by Queen Isabella upon certain men of the Royal Artillery and other branches of the service, who were employed on the North Coast of Spain, under Sir De Lacy Evans, during the Carlist War, 1835–1837.

The Royal permission for the Royal Artillery to accept and wear this decoration, is dated Whitehall, January 20th, 1843.

18.—Silver (large size.)

19.—Silver (small size.)

20.—Pewter.

There was also a silver Medal granted to officers, and a pewter one to privates, for this campaign.

Obverse.—A Maltese cross, with " St. Sebastian, 5th May, 1836," surrounded by a wreath.

Reverse.—A lion, enclosed with the collar of the Golden Fleece. " España Agradecida."

SPAIN.

MEDAL FOR BAGUR AND PALAMOS.

21.—*Obverse.* Two shields, containing the arms of Great Britain and Spain, surrounded by their colours. " Alianza Eterna."

Reverse.—" Bagur, 10 de Setiembre, Palamos, 14 de Setiembre, 1810," surrounded by " Gratitud de España a la Intrepidez Britanica."

Riband : Crimson, yellow edge.

This Medal was given to the crews of H.M.S. " Kent," " Ajax," and " Cambrian," for services in assisting the Spanish to expel the French troops from Catalonia.

TWO SICILIES.

ORDER OF ST. FERDINAND.

22.—Instituted by King Ferdinand IV., on his restoration to the throne, through the united efforts of England, Russia, and Austria, 1st April, 1800.

The Badge of the Order is a star formed of six bundles of golden rays, six Bourbon lilies intervening, and a royal crown above.

Obverse.—The figure of St. Ferdinand in regal robes ; on a dark-blue enamelled circle, " Fidei et Merito."

Reverse.—" Ferd. IV., Inst. Anno 1800."

Riband : Blue, with red edge.

TURKEY.

SULTAN'S GOLD MEDAL FOR EGYPT.

Obverse.—A crescent and star of eight points, surrounded by an ornamented border.

Reverse.—The Sultan's cypher, surrounded by an ornamented border, with the date, 1801.

Attached to the breast by a small gold hook and chain.

This Medal was given in three different sizes to the superior officers of both the Army and Navy engaged in this war.

23.—Large size.

24.—Medium size.

25.—Small size.   This Medal belonged to Lieutenant Parker, R.N.

Case XII., No. 37.

26.—Ditto, ditto.

MEDAL FOR SIEGE OF ACRE.

1 8 4 0 .

Obverse.—The Sultan's cypher encircled by a laurel wreath.

Reverse.—The Fortress of Acre, over which the Turkish standard is displayed, surrounded by six stars ; below, a Turkish inscription.

Riband : Crimson, white edge.

This Medal was given in gold to admirals, commodores, and field officers ; in silver to officers ; and in bronze to sailors and marines of the squadron commanded by Sir Robert Stopford.

London Gazette, November 17th, 1840.

27.—Gold.   This Medal belonged to Sir C. F. Smith.

Case XII., No. 34.

28.—Silver.

29.—Bronze.

MEDAL FOR GENERAL SERVICE.

30.—*Obverse.* The Sultan's cypher, within a circle, on either side of which are flags and laurel branches, surmounted by a crescent and star.

Reverse.—A large star of twelve points, in the centre of which is a smaller one of six ; below, a scroll, with Persian inscription, " Mischani Iftikhar," the decoration or mark of honour or glory, and three smaller stars.

Riband : Crimson, green edge.

This Medal was instituted by Sultan Mahomed II., in 1831, and was given in gold to officers, and silver to non-commissioned officers and men.

Sixteen men of the 10th Company Royal Engineers, under the command of Colonel, now Sir John, L. Simmons, G.C.B., received it for services at Silistria, Oltenitza, and the Battle of Mokan and the passage of the Danube at Georgevo.   Also the officers and thirty men of a gunboat, under the command of Commander H. Carr-Glyn, for services between the 7th July and 19th August, 1854, at the Sulina Mouth of the Danube.—*London Gazette, August 31st, 1855.*

SILISTRIA.

31.—*Obverse.*—The Sultan's cypher within a laurel wreath.
Reverse.—Fortress of Silistria, date Hegira, 1271 (1854).
Riband: Red, green edge.

This medal was given to Sir L. Simmons and a few English officers, viz., Butler, Nasmyth, Ballard, Ogilvy, Cannon and Hind.

KARS.

32.—*Obverse.*—Sultan's cypher within a wreath.
Reverse.—The fortress of Kars.   Hegira, 1272 (1856).
Riband: Red, green edge.

Extract from a letter from Sir F. Williams:—" I and all my officers, and an Artillery servant, received the Silver Medal."

MEDAL FOR CRIMEA.

Obverse.—The Sultan's cypher; " Crimea " in Turkish, and the year of the Hegira, 1271, within a wreath.
Reverse.—A map of the Crimea spread out over a gun; military and naval arms; the flags of Turkey, England, France and Sardinia, with the word " Crimea, 1855."
Riband: Red, green edge.

There is a variation in the arrangement of flags in these medals.

1.—English flag in centre, next to Turkish, " Crimea, 1855."
2.—Sardinian flag, next to the Turkish, " La Crimea, 1855."
3.—French flag next to the Turkish, with " La Crimée, 1855."

According to the nation for whom they were intended, owing to the wreck of a ship conveying the English Medals, there was not a sufficient number, so we find English soldiers in possession of those originally intended for the other countries.

This medal was given to all the troops in the Crimea.

33.—William Foden, 3286 ...	...	72nd Highlanders.
34.—Young, 5002	...	3rd Bat. Gren. Guards.
35.—		
36.—R. Nutting, 6160 ...	...	Grenadier Guards.
		Case XII., No.
37.—S. Lichfield	...	33rd Regiment.
		Case XII., No. 15.
38.—W. Campbell	...	79th Foot.
		Case XII., No. 25.
39.—Jas. Munro, Sergeant ...	...	93rd Highlanders.
		Case XII., No. 28.
40.—W. Hepburn	...	2nd Bat. Rifle Brigade.
		Case XII., No. 30.
41.—G. Symons, Sergeant ...	...	Royal Artillery.
		Case XII., No. 31.
42.—G. Wilkinson, Gunner & Driver		
		Case XII., No. 33.
43.—R. Lewis, Sergeant ...	...	Royal Engineers.
		Case XII., No. 35.

MEDJIDIE.

A star of eight points, between which are seven small crescents and stars of five points ; in the centre the Sultans's cypher, surrounded by a red enamelled border, containing, in Persian, the words, " Zeal, Devotion, Fidelity." It is worn suspended from a red enamelled crescent and star.

Riband : Red, green edge.

This Order was instituted by Sultan Abdid Medjid, in 1852, and given in five classes, according to the rank of the recipients, to both services for the campaign in the Crimea, 1854–1855.

44.—

45.—

CASE XII.

LIST OF OFFICERS AND SOLDIERS WHO RECEIVED
MORE THAN ONE DECORATION, ARRANGED
IN REGIMENTAL ORDER.

1st OR ROYAL DRAGOONS.

1.—War Medal. Clasps: "Fuentes D'Onor," "Vittoria," and "Toulouse."
Waterloo Medal.

JOHN DUEN.

3rd LIGHT DRAGOONS.

2.—Cabul, 1842. Spelt "Cabvl."
The late Mr. Wyon said that there were only fifteen medals with Cabul spelt in this way issued.
Punjâb Medal without clasp.

GEORGE ALBURY.

9th LANCERS.

3.—Regimental Medal from officers.
Regimental Medal from non-commissioned officers.
Long service and good conduct (William IV.)

THOMAS GODDING, Regimental Sergeant-Major.

11th LIGHT DRAGOONS.

4.—War Medal. Clasp, "Salamanca."
Army of India Medal. Clasp, "Bhurtpore."

J. BRADSHAW, Troop Sergeant-Major.

14TH LIGHT DRAGOONS.

5.—War Medal. Clasps: " Talavera," " Busaco," " Fuentes D'Onor," " Badajoz," " Salamanca," " Vittoria," " Pyrenees," " Nivelle," " Nive," " Orthes," " Toulouse."
Regimental Medal.

WILLIAM HANLEY, Troop Sergeant-Major.

16TH LANCERS.

6.—Ghuznee Medal.
Sutlej. Clasp: " Sobraon."
Punjâb: " Chilianwala, Goojerat."
Maharajpore Star.
H.E.I.C. " Long service and good conduct."
V. B. CULLEN, Sergeant-Major   ...   8th Bengal Lt. Cavalry.
Sergeant-Major Cullen served in the 16th Lancers, and obtained the Ghuznee, Sutlej, and Maharajpore medals, while in that regiment. On the return of the 16th to England, he was transferred to the 8th Bengal Light Cavalry, and received the Punjâb and H.E.I.C. medal for long service. He became Sergeant-Major.

16TH LANCERS.

7.—Maharajpore Star.
Sutlej Medal for " Aliwal; " clasp, " Sobraon."

THOMAS JOHNSON, Sergeant.

GRENADIER GUARDS.

8.—Crimea Medal. Clasps, " Alma, Balaklava, Inkerman, Sebastopol."

Turkish Medal.

Long service and good conduct.

R. NUTTING, 6160, 3rd Battalion.

This man was wounded both at Alma and Inkerman. He was transfered to the 1st battalion when they went to Canada in 1862. I, as acting Adjutant, gave him his Long Service Medal on parade at Chelsea Barracks in the summer of 1878. He died a very few weeks after in the regimental hospital, and his medals were sent to me by his brother.

2ND FOOT.

9.—Ghuznee Medal. A clasp has been added to this medal. " Campaign 1839 ; at the storming of Ghuznee, 23rd July ; at the capture of Kelat, Nov. 13th. Wm. Rofe, 2nd or Queen's Royal Regiment."

Regimental medal for 10 years.

Long service and good conduct.

WM. ROFE, Corporal.

9TH FOOT.

10.—Cabul, 1842.

Long service and good conduct. 1847. Regimental Medal.

WILLIAM BLEVINS, Sergeant.

13TH FOOT.

11.—Ghuznee Medal.

Jellalabad. (Mural crown.)

Cabul, 1842.

Regimental Medal for seven years.

J. ROUND.

12.—Ghuznee Medal, with the following clasps added:—Five Forts of Jootandurrah; Fort of Joolgar; Night attack at Baboo Khoshghur; Capture of Kardarrah; and Overthrow of Dost Mahomed Khan, 1840.

Jellalabad. Flying Victory. With clasps:—Koora Cabool Pass; Tazeen Pass; Jugdulluck Pass; Seizure of Jellalabad; and Sorties on the 11th November and 1st December, 1841.

Cabul, with clasps:—Sortie, 1st April; General action, 7th April; Jugdulluck Pass; Tazeen Pass; Re-capture of Cabool; and Re-capture of Prisoners, 1842.

JOHN WILSON, Corporal.

The clasps on these Medals were given to him by the officers of the Regiment. He was fifty-two times engaged with the enemy in India. He was field bugler to Sir G. Pollock, and greatly distinguished himself. He died in 1877. His father was in the 13th, and his son is now in the same corps in Zululand.

26TH FOOT.

13.—War Medal. Clasp: " Corunna."

Regimental Medal.

E. KENNEY, Sergeant.

31st AND 32nd FOOT.

14.—Cabul, 1842.
Sutlej Medal for Moodkee.   Clasps: " Ferozeshuhur, Aliwal,
Sobraon."
Punjáb Medal.   Clasps: " Goojerat, Mooltan."

A. Smith, Colour-Sergeant   ...   32nd Foot.

Sergeant Smith got the first two medals when serving in the 31st.
On the return of that regiment to England, he volunteered into the
32nd, and served in the Punjáb Campaign.

33rd FOOT.

15.—Crimea.   Clasp: " Sebastopol."
Turkish.
Abyssinia.

S. Lichfield.

42nd FOOT.

16.—War Medal. Clasps: " Salamanca, Pyrenees, Nivelle, Nive,
Orthes."
Waterloo.

W. Murphy, Sergeant.

52nd FOOT.

17.—War Medal.   Clasps: " Corunna, Busaco, Fuentes D'Onor,
Ciudad Rodrigo, Badajoz, Salamanca, Vittoria, Pyrenees, Nivelle,
Nive, Orthes, Toulouse."
Waterloo.
Regimental Medal for Assault of Badajoz.

Benjamin Houseley, Corporal   ...   1st Battalion.

18.—War Medal. Clasps : " Corunna, Busaco, Fuentes D'Onor, Ciudad Rodrigo, Badajoz, Salamanca, Vittoria, Nivelle, Nive, Orthes, Toulouse."
Waterloo.

JOSEPH GREEN.

51st FOOT.

19.—War Medal. Clasps : " Fuentes D'Onor, Salamanca, Pyrenees, St. Sebastian, Nivelle, Orthes."
Waterloo.
Long service and good conduct. Will. IV.

J. GREENWOOD, Sergeant.

59th FOOT.

20.—War Medal. Clasps : " Vittoria, St. Sebastian, Nive."
Waterloo.
Army of India. Clasp : " Bhurtpore."

J. LORD, Sergeant.

71st FOOT.

21.—War Medal. Clasps : " Fuentes D'Onor, Vittoria, Pyrenees, Nive, Orthes."
Waterloo.

ALEXANDER GRAHAM.

78th FOOT.

22.—War Medal. Clasp : " Java."
Army of India Medal. Clasps " Argaum, Assaye."

D. McLeod, Sergeant.
23.—General Service Medal. Clasp : " Persia."
India Mutiny. No clasp.

J. McKay.

79TH FOOT.

24.—War Medal. Clasps: " Corunna, Busaco, Fuentes D'Onor, Salamanca."
Waterloo.
Regimental Medal.

HUGH. BANNERMAN, Sergeant.

24b.—War Medal. Clasps: " Nivelle, Nive, Toulouse."
Regimental Medal.

GEORGE McKAY.

79TH FOOT.

25.—Crimea. Clasps : " Alma, Balaklava, Sebastopol."
Sardinian.
Turkish.

W. CAMPBELL.

88TH FOOT.

26.—War Medal. Clasps : " Talavera, Busaco, Fuentes D'Onor, Badajoz, Salamanca, Vittoria, Pyrenees, Nivelle, Nive, Orthes, Toulouse."
Regimental Medal, 2nd Class.

JAMES TRACEY.

91ST FOOT.

27.—War Medal. Clasps : " Roleia, Vimiera, Corunna."
Waterloo.

DAVID DEWAR.

93RD FOOT.

28.—Crimea.   Clasp: "Sebastopol."
Turkish.
India Mutiny.   Clasp: "Relief of Lucknow."
Victoria Cross.

JAMES MUNRO, Sergeant.

95TH FOOT.

29.—War Medal.   Clasps: "Salamanca, Vittoria, Pyrenees."
Pyrenees.
Waterloo.

JOSEPH HACKETT.

RIFLE BRIGADE.

30.—Crimea.   Clasps: "Alma, Inkerman, Sebastopol."
Turkish.
India Mutiny.   Clasp: "Lucknow."
   WILLIAM HEPBURN   ...   . 2nd Battalion.

ROYAL ARTILLERY.

31.—Crimea.   Clasps: "Inkerman, Sebastopol."
Turkish.
Sardinian.
Victoria Cross.
            G. SYMONS, Sergeant.
32.—War Medal.   Clasps: "Vittoria, St. Sebastian."
Waterloo.
   W. HOGG, Corporal...   ...   ... Royal Horse Artillery.
Crimea.   Clasps: "Alma, Inkerman, Sebastopol."
Turkish.
Distinguished conduct in the field.
         G. WILKINSON, Gunner and Driver.

ROYAL ENGINEERS.

34.—Field officers' Gold Medal.   " St. Sebastian."
Clasp: " Vittoria."
War Medal.   Clasp: " Martinique."
Knight Commander Charles III. of Spain.
Cross of St. Fernando, Spain.
Naval War Medal.   Clasp: " Syria."
Sultan's Gold Medal for " Acre."
Knight Commander of the Bath.

Sir C. F. SMITH, Lieut.-General.

ROYAL ENGINEERS.

35.—Crimea.   Clasp: " Sebastopol."
Turkish.
India Mutiny.   Clasp: " Lucknow."
2nd China.   Clasps: " Taku Forts, 1860 ; Pekin, 1860."

RICHARD LEWIS, Sergeant   ...   8th Company.

SAPPERS AND MINERS.

36.—India Mutiny Medal.   No Clasp.
China Medal.   Clasp: " Taku Forts, 1860."
H. E. I. C. Long service and good conduct.

H. A. GILBERT, Sergeant-Major.

ROYAL NAVY.

37.—Naval War Medal.   Clasp: " Egypt."
Naval.   Clasps: " 14th March, 1795. St. Vincent, Nile."
Small Gold Medal for Egypt: " Sultan, Selim III."
WILLIAM PARKER, Lieutenant.
38.—South Africa, H.M.S. Castor.
Crimea.   Clasps: " Inkerman, Sebastopol."
Turkish, H.M.S. Bellerophon.
Baltic, H.M.S. Pembroke.
JOSEPH WARD, A.B.
39.—War Medal.   Clasp: " Syria."
Long service, good conduct (Victoria).
    T. LEACH, Master-at-Arms ...   ... H.M.S. Victoria   and
Albert, 22 years.

ROYAL MARINES.

40.—Crimea.   Clasps: " Balaklava, Sebastopol."
Turkish.
Long service and good conduct; 21 years.
       G. VINEY   ...   . 20th Company.
41.—Arctic Medal.
Baltic.
Meritorious service.
Long service and good conduct; 32 years.
    J. A. BUTE, Sergeant ...   ... 30th Company.

ROYAL MARINE ARTILLERY.

42.—Crimea.   Clasps: " Inkerman, Sebastopol."
Turkish.
French Legion of Honour.

G. YULE, Sergeant.

29TH FOOT.

43.—War Medal.   Clasps: " Roleia, Vimiera, Talavera, Albuhera."
Naval War Medal.   Clasp, " 1st June, 1794."

THOMAS ROBSON.

FINIS.